SpringerBriefs in Environmental Science

SpringerBriefs in Environmental Science present concise summaries of cutting-edge research and practical applications across a wide spectrum of environmental fields, with fast turnaround time to publication. Featuring compact volumes of 50 to 125 pages, the series covers a range of content from professional to academic. Monographs of new material are considered for the SpringerBriefs in Environmental Science series.

Typical topics might include: a timely report of state-of-the-art analytical techniques, a bridge between new research results, as published in journal articles and a contextual literature review, a snapshot of a hot or emerging topic, an in-depth case study or technical example, a presentation of core concepts that students must understand in order to make independent contributions, best practices or protocols to be followed, a series of short case studies/debates highlighting a specific angle.

SpringerBriefs in Environmental Science allow authors to present their ideas and readers to absorb them with minimal time investment. Both solicited and unsolicited manuscripts are considered for publication.

More information about this series at http://www.springer.com/series/8868

Tanu Jindal

Emerging Issues in Ecology and Environmental Science

Case Studies from India

 Springer

Tanu Jindal
Amity Institute for Environmental Toxicology
Safety and Management (AIETSM)
and Amity Institute of Environmental Sciences (AIES)
Amity University
Noida, UP, India

ISSN 2191-5547   ISSN 2191-5555   (electronic)
SpringerBriefs in Environmental Science
ISBN 978-3-319-99397-3   ISBN 978-3-319-99398-0   (eBook)
https://doi.org/10.1007/978-3-319-99398-0

Library of Congress Control Number: 2018954249

This Springer imprint is published by the registered company Springer Nature Switzerland AG
The registered company address is: Gewerbestrasse 11, 6330 Cham, Switzerland

Foreword

Ecological crisis, or environmental deterioration, is one of the most formidable dangers facing mankind today. Unfortunately, injudicious pursuit of technological and industrial progress has created multi-directional environmental issues, bringing unforeseen hindrances in our efforts towards improving the welfare of mankind.

Everyone is a stakeholder, as we are all inhabitants of this one and only mother Earth. Each person can contribute something to address environmental pollution mitigation measures. Environmental protection means caring for our resources and subsequently for ourselves and ensuring a sustainable future for generations to come so that they will have a better environment.

It is our duty to accept personal responsibility for the success of environmental protection programs in our respective communities by cooperating and actively participating in making the atmosphere pollution free. Although, on an individual basis, we can help combat pollution in our own immediate environment. Efficient control can be best institutionalized through legislation. Thus, most countries have already addressed the issue by passing some form of pollution prevention measures.

Averting the onset of pollution in any area, be it air, water or land, could be the simplest preventive solution to the problem. This calls for a conscientious effort to adopt good practices and habits by people, the passage and the proper implementation of appropriate government laws and strict compliance, especially in industrial sectors.

Evidence shows that a number of chemicals that may be released into the air or water cause adverse health effects. The associated burden of disease can be substantial, and investment in research on health effects and interventions in specific populations and exposure situations is important for the development of control strategies. Pollution control is therefore an important component of disease control, and health professionals and authorities need to develop partnerships with other sectors to identify and implement priority interventions.

There is an urgent need to realize the severity of the consequences of environmental pollution and to engage in exerting efforts to make progress in technological and industrial aspects and, at the same time, apply innovative strategies towards combating pollution.

The chapters in this book hold many lessons. The goal is to place some of the many environmental events in a context in which they can be scrutinized objectively, systematically and passionately.

I sincerely believe that this book serves as an excellent medium for readers and enables them to explore new approaches and new outcome-based strategies.

Ashok K. Chauhan
Ritnand Balved Education Foundation
New Delhi, India

Preface

Environmental pollution is a growing menace to the human race as well as to all ecosystems on our planet. Deterioration of the environment is the world's biggest present challenge, and environmental problems such as global warming, ozone layer depletion and greenhouse effects threaten not only mankind's existence but also every living being.

Industrialization has grown at the expense of the environment, which in turn has impacted social health. Unmindful exploitation of natural resources is further causing degradation of our ecosystems.

The topics included within this book give a deeper knowledge into the correlation between abiotic stress and isoprene emission of selected plant species, temporal mount in air pollutants during specific festivals and firecracker burstings, atmospheric electrical conductivity measurements during monsoon period at a semi-urban tropical station of Northern India, activated carbon adsorbents in the treatment of wastewater, microplastics and their unsafe pathway from the aquatic environment, catalytic technology for treatment of non-biodegradable dye-polluted wastewater and dairy industry wastewater characteristics and treatment possibilities.

All topics are linked to the environment and technologies to combat pollution. Environmental pollution has repercussions on economic growth as well. Efforts need to be directed at local, regional, national and global levels to counter this problem. There is an imminent need to manage natural resources and wastes to ensure a clean and healthy environment for our future generations. Environmental education can be a potential tool in this noble cause and would ensure sustainability through awareness.

The environment has seen major action-oriented resolutions in the era of sustainability. It has been increasingly demonstrated that the future for coming generations can be secured only if environmental balance is maintained.

The book will provide an understanding to readers on various perspectives of current environmental problems in the global scenario, and pragmatic recommendations on environmental safety and public health.

Featuring a collection of informative and descriptive chapters, this book is essential reading for environmental scientists and people from related fields, as it addresses many environmental issues and practical solutions for societal benefit.

We would like to express our deepest appreciation to the scientists, faculties and research scholars who have contributed their chapters for the development of the book. In particular, Dr. Pallavi Saxena, Dr. Abhinav Garg, Dr. Adarsh Kumar, Dr. Ramanathan, Dr. R.B. Lal, Dr. Pachwarya, Ms. Guncha Sharma, Dr. Chirashree Ghosh and Dr. Rachana Singh played an important role for the book to come out in its present form. Several anonymous reviewers provided helpful comments that improved the presentation of the material in the book.

Noida, UP, India   Tanu Jindal

Contents

Atmospheric Electrical Conductivity Measurements During Monsoon Period at a Semi-Urban Tropical Station of Northern India

Adarsh Kumar

Abstract Measurements of atmospheric electrical conductivity during monsoon period at a semi-urban tropical region of Northern India were carried out and presented in this chapter. The continuous measurements of atmospheric electrical positive conductivity along with some meteorological parameters was made with the help of atmospheric conductivity meter during the monsoon season (June to August, 2010) at Roorkee (29°52′ N, 77°53′ E, 275 m above sea level). The electrical conductivity for the 3 consecutive months was in the range of 26–20, 66–17, and 55–28 × 10⁻¹⁶ S/m respectively. The atmospheric electrical conductivity was positively correlated with wind speed, relative humidity, and rainfall, while it was found to be negatively correlated with average temperature in the monsoon period. Explanations for the variation of atmospheric electrical conductivity in the light of meteorological parameters were also discussed.

Keywords Electrical conductivity · Climate change · Monsoon · Wind speed · Air temperature

1 Introduction

The sources of atmospheric air-pollution are varied and derived from the expanding needs of transporting goods and also from industrial activities, which add a large amount of gaseous and particulate contaminants to the atmosphere (Kumar 2013a, b; Deshpande and Kamra 1992). Formation of air pollutants can be attributed to natural processes or to human activities (Harrison 2005). Depending upon their origins, pollutants can be classified as biological, chemical, or physical. Physical aspects of air pollution include the emission of particulate materials from

A. Kumar (✉)
Department of Physics, Amity Institute of Applied Sciences (AIAS),
Amity University, Noida, Uttar Pradesh, India
e-mail: akumar25@amity.edu

© The Author(s), under exclusive licence to Springer Nature Switzerland AG 2019
T. Jindal, *Emerging Issues in Ecology and Environmental Science*,
SpringerBriefs in Environmental Science, https://doi.org/10.1007/978-3-319-99398-0_1

smokestacks of industrial establishments, vehicles, or incinerators (Belov et al. 2006). Carbon particles and fly ash may enter the air in large quantities (Jain and Kundu 2004). Agriculture produces physical pollutants of the atmosphere in the form of plant fibers (Kuniyal et al. 2004). Nuclear explosions spew out particulate matter into the air. Particulate matter scatter and absorb solar radiation and reduce visibility in the atmosphere (Lata and Badrinath 2005). Effects on urban temperature and other climatic conditions depend on the properties of the particles (Rogge et al. 1993). In recent years, some experimental and theoretical studies have been done to consider the effect of air pollution on atmospheric conductivity. Paoletti and Spagnolo (1989) investigated the variations of the conductivity of lower atmosphere with the presence of small traces of pollutants (NO_x) under different meteorological conditions. Guo et al. (1996) determined the atmospheric electrical conductivity as an air pollution indicator through correlation analysis. As the normal governing conditions of human lifestyle and increasing air pollution are on an average related to each other, atmospheric electrical parameters like electrical conductivity, air earth current density, and atmospheric electric field are affected by various environmental and meteorological factors and hence do not remain in a constant steady state at all the times (Saxena et al. 2010). The atmospheric electrical phenomena can be studied in two different areas, the fair weather and the disturbed weather region (Kumar 2013a, b). The local variations in the atmospheric electrical parameters are considered to provide information on the meteorological conditions in the lower layer of the atmosphere (Saxena and Kumar 2010). The study in relation to the atmospheric electrical parameters with meteorological conditions is important because both of them are related to each other. Therefore, the characteristics of the environment such as the atmospheric electrical conductivity, aerosols, moisture content, surface wind, temperature, and relative humidity are needed to be studied in greater detail (Saxena et al. 2010). To study the relationship of atmospheric electrical conductivity with meteorology, it is important to know which meteorological parameters exhibit the maximum correlation. Therefore, simultaneous measurements of unipolar conductivity and useful meteorological parameters such as temperature, relative humidity, wind speed, and rainfall should be made and the time-dependent nature of these elements should also be taken into account. The aim of the present work is to study the experimental results of direct measurements of electrical conductivity and correlate it with some of the meteorological parameters like air temperature, relative humidity, wind speed, and rainfall. Therefore, a complete measurement of atmospheric conductivity was made at an altitude of about 12 m from the ground in the building of the Department of Physics, Indian Institute of Technology, Roorkee during the monsoon season from June to August, 2010.

2   Methodology and Instrumentation

Measurements of atmospheric conductivity were made with the help of atmospheric conductivity meter (Singh et al. 1999). A detailed description of atmospheric electrical conductivity meter is available elsewhere (Saxena et al. 2010). It consists of the sensor for ionic concentration, the electrometer amplifier, and the chart recorder. The sensor is basically a cylindrical capacitor and has two coaxial cylinders between which air is allowed to flow (Kumar 2014a, b). Out of the two cylinders, the outer cylinder was just used to capture all the incoming particles so that they may pass through the inner cylinder. The diameter of the inner cylinder was 9.5 cm while outer cylinder was made up of 10.5 cm. The intake of air was aided by an air blower with a capacity of 1500 liter per minute. When the air is drawn in, the ions of opposite polarity are attracted by the electrodes. A positive potential at the outer electrode makes electrons and negative ions move to the central electrode, thereby constituting a current. This gives the negative conductivity. Also, the negative potential on the outer electrode similarly provides us the positive conductivity. This way, the positive conductivity was recorded along with some meteorological parameters like as wind speed, temperature, and relative humidity. For the measurements, the instruments are installed on the top of a building near Roorkee (India) at a height of about 12 m from the ground, far from any type of interference (Kumar et al. 2011). Mean daily values of atmospheric electrical conductivity for the monsoon season covering June to August, 2010 were obtained by averaging electrical conductivity at every 10 min interval during the course of a particular day. However, the daily routine meteorological data for temperature, relative humidity, wind speed, and rainfall were measured by using the normal meteorological instruments.

3   Results and Discussion

The experimental daily mean variation of atmospheric electrical conductivity, wind speed, average temperature, relative humidity, and rainfall for the monsoon period (June to August) of 2010 respectively has been shown in Fig. 1a–e. The atmospheric electrical electrical conductivities for the 3 consecutive months were in the range of (25.2–19.5), (65–18), and (53–26) $\times$ 10^{-16} S/m, respectively (Figs. 2a, 3a and 4a) while the relative humidity was found to be within limits of (95–63), (96–75), and (75–58) for June, July, and August, respectively (Figs. 2b, 3b and 4b). For these three consecutive months of 2010, the mean conductivity was 21.29×10^{-16}, 40.55×10^{-16}, and 37.39×10^{-16} S/m, respectively. The observed atmospheric electrical conductivity has been positively correlated with wind speed, relative humidity, and rainfall, while it was found to be negatively correlated with average temperature in the monsoon period of 2010 (Figs. 5a–d). The average rainfalls for these 3 months were 1.4 mm, 5.6 mm, and 8.08 mm per day, respectively (Figs. 2e, 3e and 4e). However, it rarely rained in the month of June except on the last few days (Fig. 2e). It was found that the atmospheric conductivity increases with increasing monthly rainfall (Fig. 5d). It was attributed to the fact that the rain contains charged particles, which increases atmospheric conductivity near ground surface. The increased relative humidity was due to the increased moisture content of the atmosphere. Kumar (2014a), b) mentioned that the electrical conductivity of the moist air was more than that of the dry air. It is to be noted that July and August are the months of rain. Condensation of water molecules starts on aerosol particles

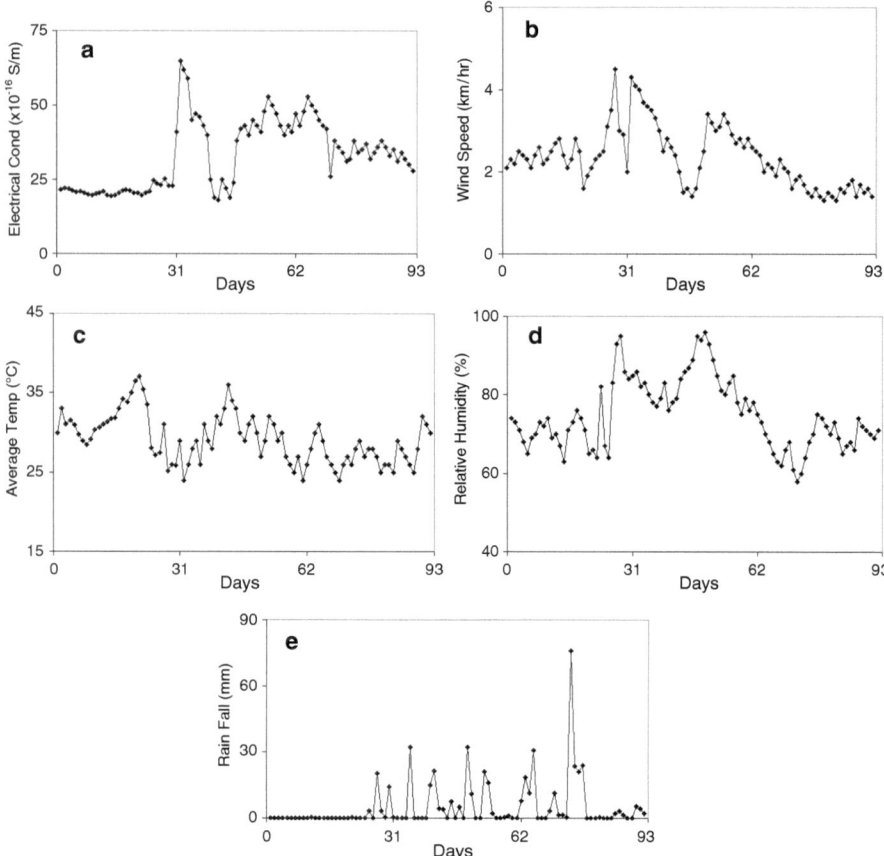

Fig. 1 (**a–e**) Variation of electrical conductivity, wind speed, average temperature, relative humidity, and rainfall during the monsoon period of 2010

(Saxena et al. 2012). The atmospheric ions which are attached to the droplets and aerosols decrease the mobility of these ions. Even when the droplets grow to become a drop and further rain, it brings the atmospheric charge carriers to the earth surface. Thus the increasing relative humidity decreases the atmospheric conductivity by reducing the mobility of charge carriers. Kumar et al. (1998) found that the changes in relative humidity were inversely related to atmospheric electrical parameters. Similar views were also expressed by Deshpande and Kamra (1992).

Figure 5b shows that during all the 3 months of monsoon, the atmospheric conductivity was negatively correlated with temperature. Kumar (2014a, b) suggested that with the onset of solar eclipse of March 7, 1980, an increment in both types of conductivity was reported. However, it is well known that the solar eclipse event is followed by a decrease in temperature. This has been significantly attributed to the increased relative humidity in the atmosphere and hence the increase in atmospheric conductivity. However, the atmospheric conditions during the monsoon period were

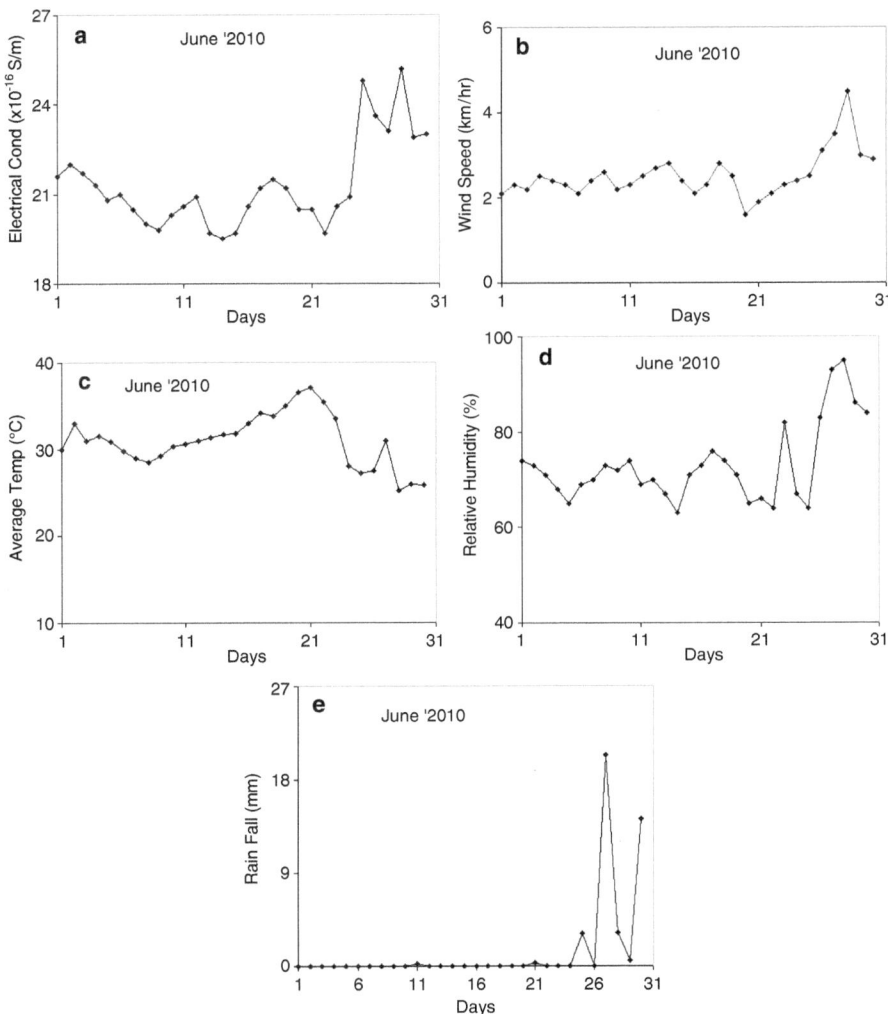

Fig. 2 (**a–e**) Variation of electrical conductivity, wind speed, average temperature, relative humidity, and rainfall during the month of June 2010

quite different. During the monsoon period, water droplets were formed in the atmosphere and thereby increasing the total aerosol concentration. An increase in air temperature vaporizes the existing droplets and a decreasing temperature creates more droplets due to increased condensation. Therefore, the increasing temperature decreases the atmospheric aerosol concentration, and thereby decreasing ionic mobility and hence decreasing the atmospheric electrical conductivity. The present results on positive atmospheric conductivity and wind speed show a positive correlation for the 3 months of June, July, and August respectively and an enhancement was noted with increasing wind speed. This enhancement of positive electrical con-

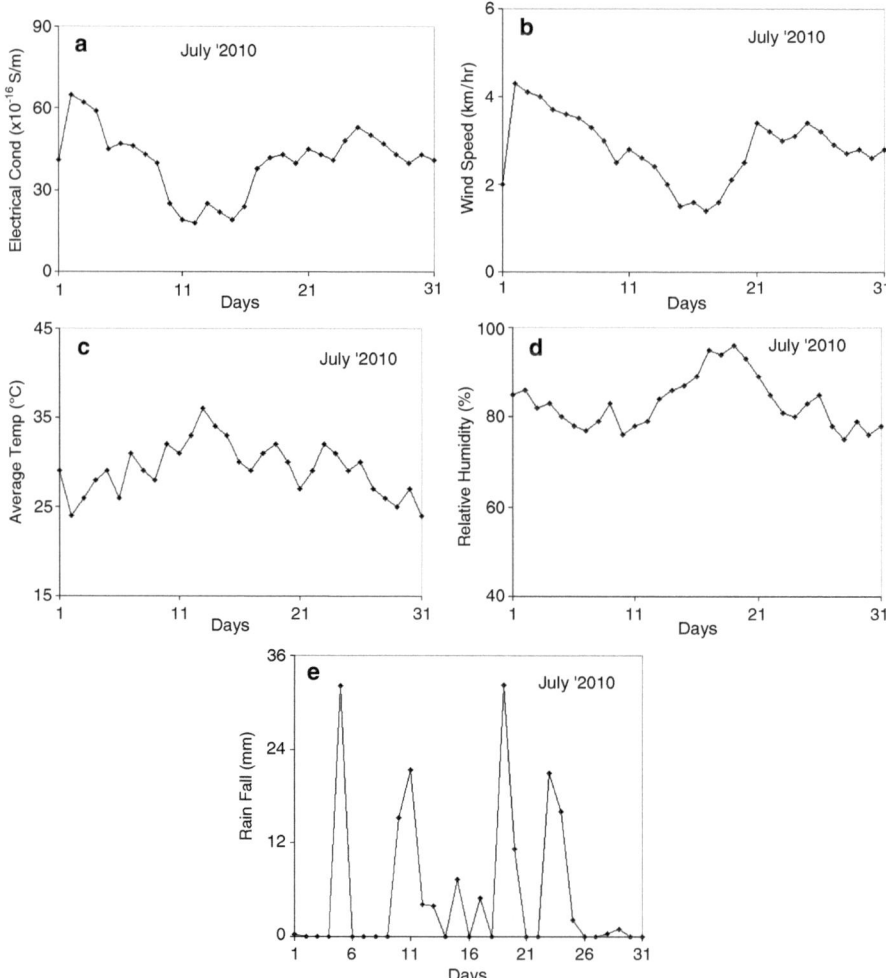

Fig. 3 (**a–e**) Variation of electrical conductivity, wind speed, average temperature, relative humidity, and rainfall during the month of July 2010

ductivity with wind speed is in close agreement with the work of Harrison (2005) where he reported that wind influences conductivity mostly in the short span of 3 months range during monsoon period. The high wind speed removes the larger atmospheric particles faster and thereby increasing the mobility of smaller sized charged particles. The increased mobility increases the atmospheric electrical conductivity. Thus the increasing wind speed increases the air conductivity of the atmosphere.

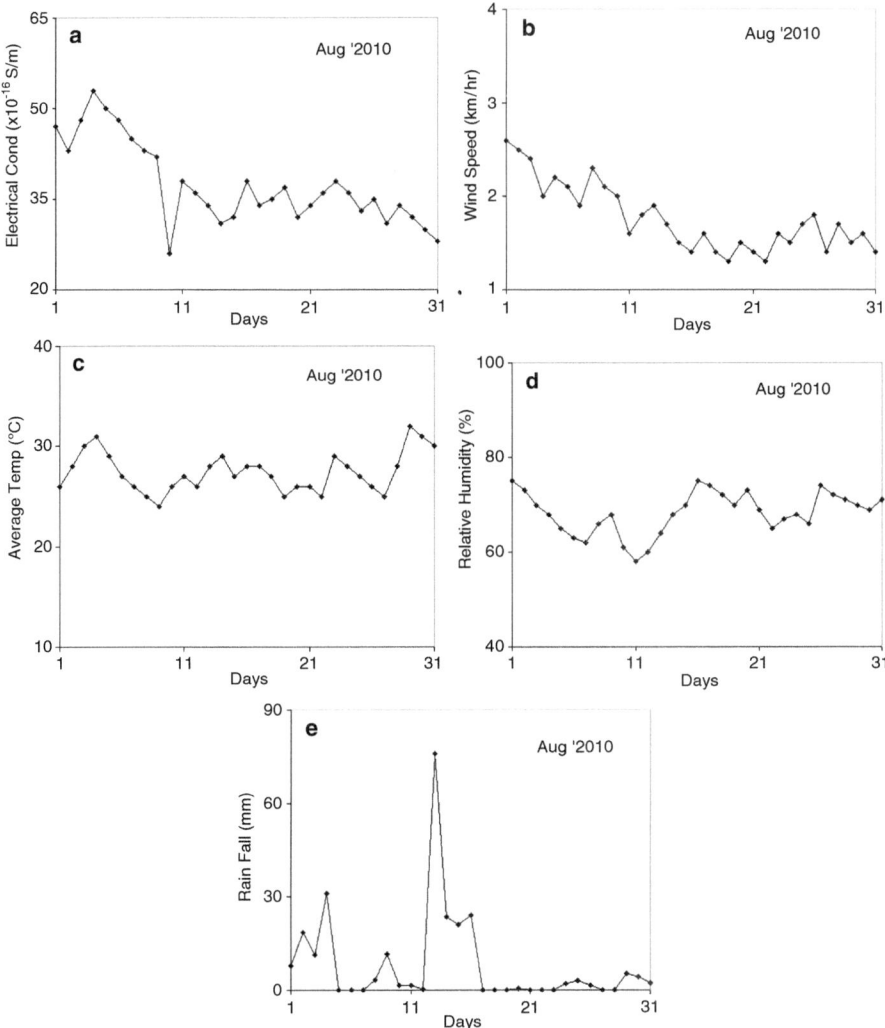

Fig. 4 (**a–e**) Variation of electrical conductivity, wind speed, average temperature, relative humidity, and rainfall during the month of August 2010

4 Conclusions

The present work studies the variations of experimentally measured positive conductivity and meteorological parameters such as relative humidity, temperature, wind speed, and rain for the 3 months of monsoon period (June to August 2010). Following are the conclusions of the present work:

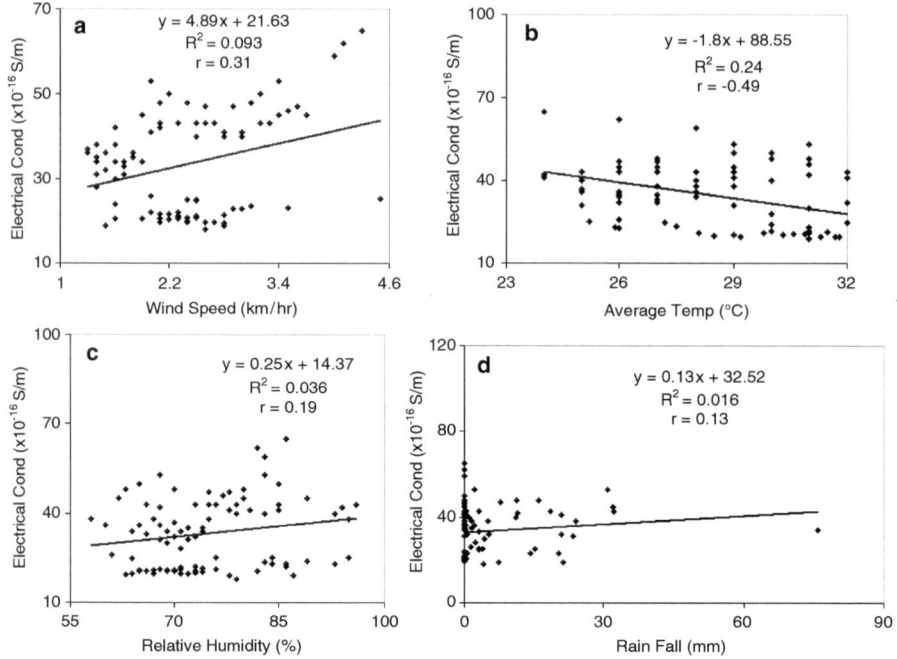

Fig. 5 (**a–d**) Correlation of electrical conductivity with wind speed, average temperature, relative humidity, and rainfall

1. Atmospheric electrical conductivity was positively correlated with wind speed, relative humidity, and rainfall, while it was found to be negatively correlated with average temperature in the monsoon period.
2. The short range analysis of present atmospheric data shows the wind as an important factor that modifies the behavior of electrical conductivity.
3. Whereas the measurements were taken at a semi-urban location of India but the findings were expected to be valid for all the sub tropical regions also.
4. The important findings of the present chapter hint at the short range variations in surface atmospheric electrical parameters by incorporating the meteorological parameters.

Acknowledgement The corresponding author is grateful to Prof. A.L. Verma, Advisor (Science & Technology-Research), Amity University, Noida, India for providing motivation and necessary computational facilities towards this work. Special thanks are also highly due to Prof. J. Rai, Indian Institute of Technology, Roorkee, India for engaging in fruitful discussions towards completing the present work.

References

Belov AV, Gushchina RT, Obridko VN, Shelting BD, Yanke VG (2006) Long-term variations of galactic cosmic rays in the past and future from observations of various solar activity characteristics. J Atmos Sol Terr Phys 68(11):1161–1166

Deshpande CG, Kamra AK (1992) Short time variations in atmospheric electrical parameters. J Atmos Sol Terr Phys 54(11–12):1413–1420

Guo Y, Barthakur NN, Bhartendu S (1996) Using atmospheric electrical conductivity as an urban air pollution indicator. J Geophys Res 101(D4):9197–9203

Harrison RG (2005) Columnar resistance changes in urban air. J Atmos Sol Terr Phys 67(8–9):763–773

Jain M, Kundu N (2004) Effect of volcanic aerosols on stratospheric ozone at Kodaikanal. Indian J Radio Space Phys 33:256–259

Kumar A (2013a) Variability of aerosol optical depth and cloud parameters over North Eastern regions of India retrieved from MODIS satellite data. J Atmos Sol Terr Phys 100–101:34–49

Kumar A (2013b) Variation of atmospheric electrical conductivity during monsoon period at a semi-urban tropical station of Northern India. Indian J Phys 87:411

Kumar A (2014a) Long term (2003–2012) spatio-temporal MODIS (Terra/Aqua level 3) derived climatic variations of aerosol optical depth and cloud properties over a semi arid urban tropical region of Northern India. Atmos Environ 83:291–300

Kumar A (2014b) A comparative study on orographic and latitudinal features of global atmospheric electrical parameters over different places at three Asian countries. Indian J Phys 88(3):225–235

Kumar A, Nigam MJ, Singh AK, Nivas S, Rai J (1998) Effect of orographic features on atmospheric electrical parameters over different cities of India. Indian J Radio Space Phys 27:215–223

Kumar A, Saxena D, Yadav R (2011) Measurements of atmospheric aerosol concentration of various sizes during monsoon season at Roorkee, India. Atmos Sci Lett 12(4):345–350

Kuniyal JC, Momin GA, Rao PSP, Safai PD, Tiwari S, Ali K, Khwairakpam G (2004) Aerosols behaviour in sensitive areas of the northwestern Himalaya: a case of Kullu-Manali tourist complex, India. Indian J Radio Space Phys 33:170–175

Lata KM, Badrinath KVS (2005) Effect of aerosols on erythemal ultraviolet radiation. Indian J Radio Space Phys 34:175–178

Paoletti D, Spagnolo GS (1989) Atmospheric electricity in a rural site and its possible correlation with pollution: a preliminary study. Atmos Environ 23(7):1607–1611

Rogge WF, Mazurek MA, Hildemann LM, Cass GR, Simoneit BRT (1993) Quantification of urban organic aerosols at molecular level: identification, abundance and seasonal variation. Atmos Environ 27:1309–1330

Saxena D, Kumar A (2010) Measurements of atmospheric electrical conductivities during the total solar eclipse of 22 July, 2009. Indian J Phys 84(7):783–789

Saxena D, Yadav R, Kumar A (2010) Effect of orographic features on global atmospheric electrical parameters over 160 different places of US. Indian J Phys 84(4):383–389

Saxena D, Yadav R, Kumar A (2012) Orographic features of global atmospheric fair weather electrical parameters over different places of Sri Lanka. Sri Lankan J Phys 13(1):9–16

Singh AK, Nivas S, Kumar A, Rai J, Nigam MJ (1999) Variations in atmospheric aerosols and electrical conductivity at Roorkee during the total solar eclipse of October 1995. Indian J Radio Space Phys 28:1–10

Temporal Mount in Air Pollutants Allied with Religious Fiesta: Diwali, Festival of Lights

Abhinav Garg, Priyanka Sharma, Gufran Beig, and Chirashree Ghosh

Abstract India being a secular country is bound by many religious and cultural activities. Diwali or Deepawali is one of the major festivals which is celebrated with great heartiness by bursting firecrackers (both aerial and land). "Firework displays in Delhi, the capital of India, are known to be grand and extensive, perhaps owing to the relative affluence of the city's population" (Sarkar et al. 2010). Fireworks are one of the most unusual sources of anthropogenic pollution in atmosphere; although short-lived, these events are responsible for high concentrations of particles and gases in the atmosphere. To recognize the severity of such episodes on air quality, we undertook a study analyzing the pollutant data the day before, on the eve of Diwali, and the day after the fiesta, for 6 consecutive years, 2010–2015. Further, we compare the results by taking a reference month, the month before the celebration. Pollutant data for particulate matter (PM_{10} and $PM_{2.5}$), nitrogen dioxide (NO_2), and ozone (O_3) was collected from real-time SAFAR (MoES) monitoring station installed at Sports Stadium, University of Delhi, Delhi. To understand the health impacts and dispersion of pollutants, AQHI and Hysplit model was also computed. On average on the day of the festival the concentration of PM_{10} and $PM_{2.5}$ was found to be 6.5 and 6.8 times higher respectively in comparison to permissible limit over period of 6 years. Similarly, NO_2 and O_3 were found crossing the permissible limit by 1.6 and 1.2 times in the year 2013. Interestingly, in the year 2011 and 2014 restricted dispersion was observed, with winds moving back towards Delhi, which might have had a role in concentrating the pollution. The short-term exposure of these pollutants beyond the permissible limits can increase the probability of acute health effects, especially for the vulnerable group of population. Therefore, it is of utmost necessity to monitor the outburst of pollution during such events. Although regulating or restricting the festival would be socially unacceptable, adapting to cleaner, more environmental friendly ways to celebrate the festival is the need of the hour.

A. Garg (✉) · P. Sharma · C. Ghosh
Environmental Pollution Laboratory, Department of Environmental Studies,
University of Delhi, Delhi, India

G. Beig
Indian Institute of Tropical Meteorology, Pune, India

© The Author(s), under exclusive licence to Springer Nature Switzerland AG 2019   11
T. Jindal, *Emerging Issues in Ecology and Environmental Science*,
SpringerBriefs in Environmental Science, https://doi.org/10.1007/978-3-319-99398-0_2

Keywords Diwali · Firecracker · Criteria pollutants · PM_{10} · $PM_{2.5}$. NO_2 · O_3 · AQHI · Hysplit

1   Introduction

India is a land of different religions and cultures with different festivals; one festival which is celebrated with all zeal and vigor is Diwali or Deepawali. The festival is generally celebrated during the month of October/November every year which is marked by the bursting of firecrackers all over India. Culture and environment are interlinked with each other; cultural practices directly or indirectly impact environment, so does the impact of Diwali celebration on environment, by and large ambient air quality is compromised. Delhi which is ranked among the world's most polluted cities (WHO 2014) faces additional huge pollution load during the days of Diwali celebration. "Firework displays in Delhi, are known to be grand and extensive, perhaps owing to the relative affluence of the city's population" (Sarkar et al. 2010). Widespread use of firecrackers contributes particulate and gaseous pollutants in air such as particulate matter (PM), oxides of sulfur (SO_x), oxides of nitrogen (NOx), carbon dioxide (CO_2), and carbon monoxide (CO) (Singh et al. 2010). High concentrations of PM_{10} and $PM_{2.5}$ were found during and after Diwali day, especially in the Indo-Gangetic plain (Kumar et al. 2016).

Not only primary and secondary pollutants are released from firecrackers but also toxic metal concentration increases during Diwali. Pal et al. (2013) reported that concentration of metals such as Zn, Al, Cu, Pb, Fe, Mn, Cr, Cd, and Ni increases during Diwali day as compared to their concentration in normal days. Kumar et al. (2016) strongly considered elements like K, Ba, Sr, Cd, S, and P to be firework tracers showing higher concentrations during Diwali. Particulates adsorb trace metals and organic compounds onto their surface, and when these particles inhaled pose severe threat to human health (Ravindra et al. 2001). Colored fireworks also generates ozone at the ground level, surface ozone is harmful being a strong oxidizing agent (Attri et al. 2001).

The ambient noise level has also been reported to be 1.2 to 1.3 times higher during Diwali day as compared to normal days (Mandal et al. 2012). So this fiesta is not only causing air pollution but also causing noise pollution, taking a toll on both environment and human health.

The fine particulate matter released by bursting of crackers can pose a great risk to respiratory health and can cause problems like bronchitis, pneumonia and lung cancer (Huang et al. 2003; Vernath et al. 2004). Even, high inhalation of PM may cause tachycardia leading to decline in heart rate variability which increases the risk for sudden cardiac death and heart attack (Pal et al. 2013). The Air Quality Index (AQI) study conducted by Parkhi et al. (2016), at different areas of Delhi concluded that the AQI for $PM_{2.5}$ and PM_{10} before, during and after Diwali remains to be "Very Unhealthy." Thus, Delhi which already has stigma to be one of the most polluted cities in the world, such a celebration even worsens this condition.

The present study aims to quantify the concentration of certain criteria pollutants like PM_{10}, $PM_{2.5}$, O_3, NO_2 in and around Diwali fiesta. The dispersion of pollutants was also studied to understand how the pollutants released during Diwali celebration disperse, using a single forward trajectory Hysplit model. In the end the impact of multiple pollutants released during Diwali on air quality and human health was indexed by computing "Air Quality Health Index" (AQHI).

2 Methodology

2.1 Study Area: Delhi

Study was undertaken in National Capital Territory of Delhi, situated in the northern part of India stretching from the latitude of $28°24'17''$ to $28°53'$ and from the longitude of $76°20'37''$ to $77°20'37''$ at an altitude between 213 and 305 m above sea level. It is characterized by semi-arid climate, mainly influenced by its inland position and prevalence of continental air during most of the year, with weather fluctuating between extreme hot to extreme cold, ~40.5 °C and ~6.7 °C. The northern part of India, especially Indo-Gangetic Plain, encounters a foggy weather condition during winter months together with lower mixing heights, leading to poor dispersion and mixing with the upper boundary layer, thus rendering reduced visibility. The temperature rapidly decreases toward the end of October (Ali et al. 2004), and continental air masses enriched with pollutants pass over Delhi during winter months (Tiwari et al. 2012). In general winds are predominantly Westerly or North-Westerly during winter and Easterly and South-Easterly in monsoon months.

2.2 Data Collection and Analysis

To recognize the severity of short lived episodes on air quality, we analyzed the pollutant data, the day before, on the eve of Diwali, and the day after the fiesta, for consecutive 6 years, 2010–2015. Further, we compare the results by taking a reference month, the month before the celebration, for understanding of the background pollution concentration. Pollutant data for particulate matter—respirable particulate matter (RPM, $PM_{10})$ and fine particulate matter (FPM, $PM_{2.5}$), nitrogen dioxide (NO_2), and ozone (O_3) was collected from real time "System of Air Quality Forecast and Research" (SAFAR) Programme (Ministry of Earth Sciences (MoES)), for one of the monitoring stations installed at Sports Complex, University of Delhi, Delhi. At the SAFAR University Station, Particulate Matter (PM_{10} and $PM_{2.5}$) is continuously monitored using Beta Attenuation Monitor (BAM-1020; Met One Instruments, Inc., USA), with a resolution of 0.1 µg m^{-3} and lower detection limit of around 1 µg m^{-3}; Tropospheric or Ground Level Ozone and NO_2, are monitored using O_3 analyzer (49i; Thermo Scientific, USA) with precision of ~1 ppbv and NOx analyzer (42i; Thermo Scientific, USA) with precision of ~0.4 ppbv respectively.

All the pollutant data is continuously recorded at the 15-min time interval for 6 consecutive years (2010–2015), data obtained from continuous monitoring was computed and (a) averaged hourly, to make diurnal graph for the eve of Diwali, (b) averaged 24 hourly for the day before, the day after and on the eve of Diwali and (c) averaged monthly (1 month), for a month before the month on which Diwali falls, to understand the background concentration.

2.3   Dispersion Modeling

Atmospheric dispersion modeling is executed to understand how air pollutants disperse in the atmosphere. This is usually performed with a computer programs incorporating mathematical equations and algorithms which simulate the pollutant dispersion. The dispersion models are used to estimate the downwind ambient concentration of air pollutants released from different sources. One of the dispersion models is HYSPLIT, which we have used in our study.

HYSPLIT Model: The Hybrid Single Particle Lagrangian Integrated Trajectory Model (**HYSPLIT**) is a computer model that is used to simulate air parcel trajectories and dispersion or deposition, complex transport, chemical transformation of atmospheric pollutants. It was developed in collaboration of National Oceanic and Atmospheric Administration (NOAA) and Australia's Bureau of Meteorology. Application of the model which is been exploited in the study is the back trajectory analysis, to determine the origin of air masses and establish source–receptor relationships. The model calculation method is a hybrid between the Lagrangian approach, using a moving frame of reference for the advection and diffusion calculations as the trajectories or air parcels move from their initial location (Stein et al. 2015).

2.4   Burden of Illness from Air Pollution: Risk Communication

Health impacts associated with exposure to air pollution are mostly estimated using a single pollutant approach, like Air Quality Index (AQI) (U.S. EPA), Air Pollution Index (API, Hong Kong), but people are exposed to mixture of multiple pollutants in the air that may have individual or combined effects on human health. So, it is important to develop an exposure metrics that represent the multipollutant environmental impacts on human health. Air Quality Health Index (AQHI), is one such multipollutant index which evaluate the impact of multiple air pollutants on human health.

Air Quality Health Index (AQHI): It is a communication tool which is designed to help one make decisions to protect their health by limiting short-term exposure to air pollution and adjusting their activity levels during increased levels of air pollution. It measures the air quality in relation to human health on a scale from 1 to 10. The higher the number, the greater the health risk associated with the air quality levels, making it easy and effective communicating tool.

AQHI was developed by Stieb with his team in 2005, by estimating the concentration-response coefficients from a time-series analysis of air pollution and mortality from multiple cities. Then they applied these coefficients for the calculation of the generic index, together with various sensitivity analyses based on the application of coefficients from different models and different sources from the literature. Stieb and team further strengthened the index in 2008, by improving and testing with more concentration–response coefficients from various time series analyses.

The AQHI is calculated from the sum of the percentage added health risk (HR) of daily hospital admissions attributable to the maximum 3-h moving (rolling) average concentration of criteria air pollutants, Ozone (O_3), Nitrogen Dioxide (NO_2), and Particulate matter (Eqs. 1–5) for the eve of Diwali. The HR of each pollutant depends on its concentration and a risk factor (coefficient) which was derived from earlier studies (To et al. 2013; Stieb et al. 2008, 2005). The HR is then compared to a scale to obtain the appropriate banding of AQHI (Table 1).

$$HR = HR(NO_2) + HR(O_3) + HR(PM) \qquad (1)$$

Where, HR (PM): Added health risk from Particulate Matter, HR(O_3): Added health risk from Surface Ozone, HR(NO_2): Added health risk from Nitrogen Dioxide.

Health Risk from Respective Pollutant is calculated using the following equations:

$$HR(NO_2) = \left[\exp\left(\beta(NO_2) \times C(NO_2)\right) - 1 \right] \qquad (2)$$

$$HR(O_3) = \left[\exp\left(\beta(O_3) \times C(O_3)\right) - 1 \right] \qquad (3)$$

$$HR(PM_{10}) = \left[\exp\left(\beta(PM_{10}) \times C(PM_{10})\right) - 1 \right] \qquad (4)$$

$$HR(PM_{2.5}) = \left[\exp\left(\beta(PM_{2.5}) \times C(PM_{2.5})\right) - 1 \right] \qquad (5)$$

Where, $C(NO_2)$, $C(O_3)$, $C(PM_{10})$ and $C(PM_{2.5})$ are the 3-h moving average concentration of the respective pollutants, $\beta(NO_2)$, $\beta(O_3)$, $\beta(PM_{10})$ and $\beta(PM_{2.5})$ are regression coefficients of the respective pollutants.

Table 1 The AQHI categories and health remark

Health ri Sk	AQHI category	Health messages	
		At-risk population	General population
Low	1 to 3	**Enjoy** your usual outdoor activities	**Ideal** air quality for outdoor activities
Moderate	4 to 6	**Consider** reducing or rescheduling strenuous outdoor activities if you are experiencing symptoms	**No need to modify** your usual outdoor activities unless you experience symptoms of coughing and throat irritation
High	7 to 10	**Reduce** or reschedule strenuous outdoor activities. Children and the elderly should also take it easy	**Consider reducing** or rescheduling strenuous outdoor activities if you experience symptoms of coughing and throat irritation
Very high	Above 10 (10^+)	**Avoid** strenuous outdoor activities. Children and the elderly should avoid outdoor physical exertion	**Reduce** or reschedule strenuous outdoor activities, especially if you experience symptoms of coughing and throat irritation

Source: Abelsohn and Stieb (2011)

3   Result and Discussion

3.1   Variation in Concentration of Pollution

Particulate Matter: PM_{10} and $PM_{2.5}$

The analysis of data depicted that the concentration of Particulate matter, Respirable and Fine Particulate matter, PM_{10} and $PM_{2.5}$ respectively, had crossed the permissible limit set by NAAQS (PM_{10}–100 μg/m^3, $PM_{2.5}$–60 μg/m^3) in all the studied years, 2010–2015 (Figs. 1 and 2). In 2010, the day after Diwali, the concentration of PM_{10} and $PM_{2.5}$ was found to be 19.5 and 25.6 times higher, respectively, than the permissible limit. Over a period of 6 years, on average on the day of Diwali, PM_{10} and $PM_{2.5}$ were 6.5 and 6.8 times higher, respectively, in comparison with the permissible limit. For PM_{10} on the day of Diwali, year 2010 was the most polluted year showing concentration of 986.6 μg/m^3, whereas lowest concentration was found on 2015, 434.15 μg/m^3. For PM 2.5 on the day of Diwali, year 2010 was the most polluted year, 656.15 μg/m^3; on the contrary year 2014 had observed the lowest concentration, 235.83 μg/m^3 (Figs. 1 and 2).

Particle concentration a month before, on the day before, on the eve, and the day after Diwali showed that the concentration was lowest a month before Diwali but increased on the eve or the day after Diwali in all the selected years (2010–2015). There was steep decline in the particulate concentration after 2010, but steady decreasing trend was observed particularly after 2013.

Several studies had also reported high concentration of PM during Diwali Days. A study conducted by Tiwari et al. (2012) reported high concentration of PM_{10}, $PM_{2.5}$ and PM_1 during and the day after Diwali. Also, in the same study lower PM

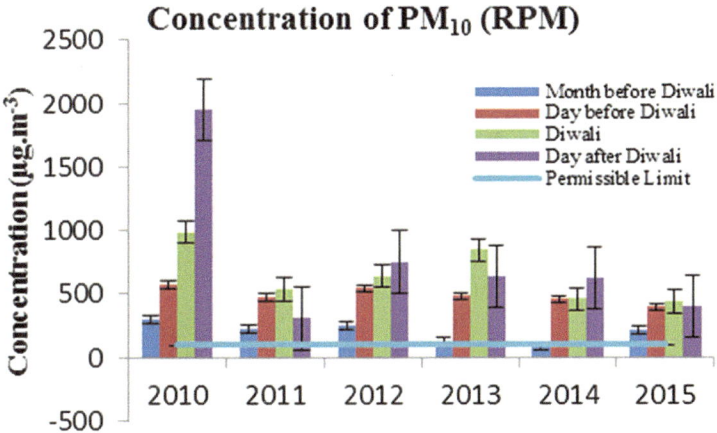

Fig. 1 Variation in concentration of respirable particulate matter (RPM, PM$_{10}$)

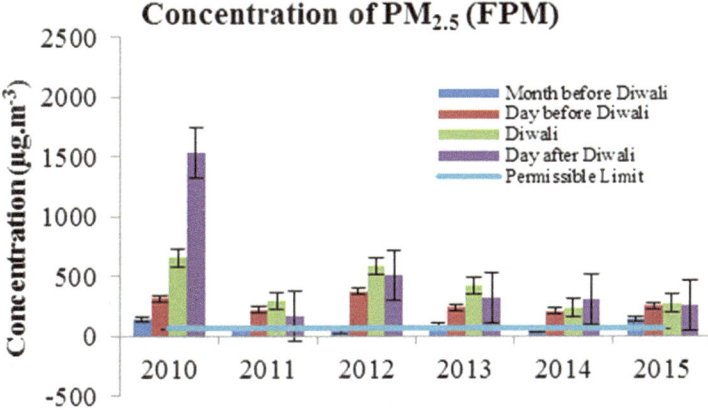

Fig. 2 Variation in concentration of fine particulate matter (FPM, PM$_{2.5}$)

concentration was reported on the festive day in year 2008 as compared to 2007 and this was attributed to increased mixing height, temperature, and wind speed in the year 2008. Thus, pointing that different year can show different trend depending on the prevailing meteorological conditions. Overall, subsequent decrease in the PM concentration after 2010 can be due to the increased awareness in citizens regarding harmful effects of bursting firecrackers. Peshin et al. (2017) also reported decreased concentration of pollutant in year 2014 and 2015 and explained that due to increase in awareness campaign and cost of firework such results have been observed.

High concentration of particulates in inhaled air has high possibility to reach deep in lungs, especially of size (diameter) smaller than 10 μm. Deposition of particles into lungs is dangerous, because they can enrich complex mixture of toxic and hazardous pollutants released from fireworks on their surface (Ravindra et al. 2003),

Table 2 Epidemiological review showing combined effect estimates of daily mean of PM

Health impacts	Percentage change in health indicator with each 10 μgm^{-3} increase in PM$_{10}$ concentration
Increase in daily mortality	
Respiratory deaths	3.4
Cardiovascular deaths	1.4
Increase in hospital visit (respiratory diagnoses only)	
Admissions	1.4
Emergency department visits	0.9
Exacerbation of asthma	
Asthmatic attacks	3.0
Bronchodilator use	12.2
Emergency department visit	3.4
Hospital admissions	1.9
Increase in reported respiratory symptom	
Lower respiratory	3.0
Upper respiratory	0.7
Cough	2.5
Decrease in lung functioning	
Forced expiratory	0.15
Peak expiratory flow	0.008

Source: COEHA (1996)

which when reach to the alveoli gets dissolved and enters into the blood stream (Hoet et al. 2004).

Numerous Epidemiological studies for acute adverse effects following exposure to particulates estimates the effects as percentage increase in mortality associated with each incremental increase of PM$_{10}$ by 10 μg m^{-3} (Stieb et al. 2008; Schwartz 1993; Dockery et al. 1993; Pope et al. 1992) (Table 2).

Gaseous Pollution: Nitrogen Dioxide and Surface Ozone

During the study period both the gaseous pollutants, nitrogen dioxide (NO$_2$) and surface ozone (O$_3$), majorly showed concentration within the allowable permissible limit set by NAAQS, 80 μg/m^3 and 100 μg/m^3, respectively, except for the year 2010 and 2013 for NO$_2$, and year 2013 for O$_3$ (Figs. 3 and 4). NO$_2$ concentration in 2010 and 2013 crossed the permissible limit by 3.5 and 1.6 times respectively on the day of Diwali. concentration was 1.2 times the permissible limit, the day before the main event of Diwali, 2013.

Gaseous pollutants showed a different trend than particulate pollution; NO$_2$ was highest in 2010 (283.37 μg/m^3) followed by year 2013 (131.67 μg/m^3) on the day of

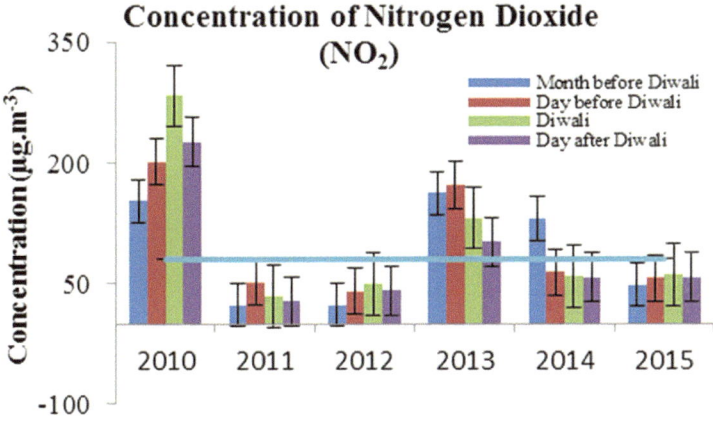

Fig. 3 Variation in concentration of nitrogen dioxide (NO₂)

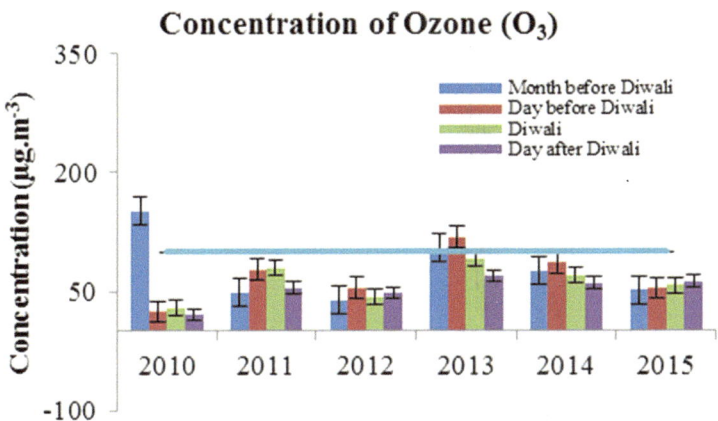

Fig. 4 Variation in concentration of ozone (O₃)

Diwali. However, no clear effect of Diwali celebration (cracker bursting) was observed on NO_2 concentration except for the year 2010 and 2012. In the case of O_3 highest concentration was observed in 2013, 91.37 μg/m³ and lowest concentration was observed in 2010, 28.93 μg/m³ on the day of Diwali. In most of the years (2011–2014) concentration was observed to be high on the day before Diwali. Thakur and Bhatia (2015) also observed similar results where NO_2 and O_3 concentrations were higher during pre-Diwali time as compared to Diwali eve and pointed out that not just Diwali celebration has an influence on concentrating gaseous pollutants but other dynamic natural phenomenon and anthropogenic activities like weather phenomena and vehicular emission also play an important role in governing the concentrations.

Fig. 5 Diurnal variation in concentration of respirable particulate matter (RPM, PM_{10})

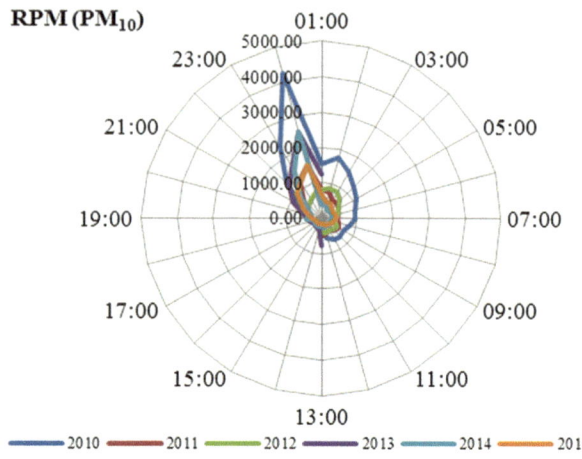

Fig. 6 Diurnal variation in concentration of fine particulate matter (FPM, $PM_{2.5}$

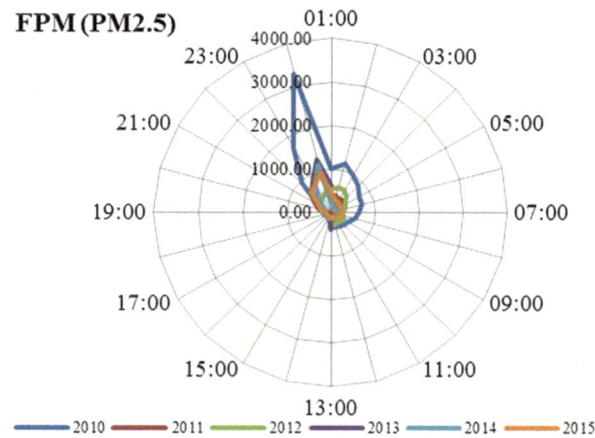

3.2 Diurnal Variation in Pollution Concentration: Diwali Eve

Hourly graphs (00:00 to 24:00 h) were plotted for each pollutant, particulate matter (PM_{10} and $PM_{2.5}$), nitrogen dioxide (NO_2), and ozone (O_3) for 6 years, 2010–2015 for the day of Diwali (Figs. 5, 6, 7, and 8). The diurnal variation in pollution concentration showed that all the pollutants showed peak concentration during night of Diwali from 23:00 to 24:00 h.

NO_2 showed higher concentration during night increasing from 11:00 PM to 3:00 AM the other day (Fig. 6). Ozone interestingly recorded a dumbbell shaped pattern (in radar graph, Fig. 7) where concentration in a day increased twice, first during afternoon from 1:00 PM to 3:00 PM and second time, during night from 11:00 PM to 2:00 AM. Increase in concentration of ozone during afternoon is attributed to its life cycle which follows the formation of tropospheric ozone by photochemical process. However, high ozone concentration during night was also reported by Attri et al. (2001) which explained that ozone can also be produced from

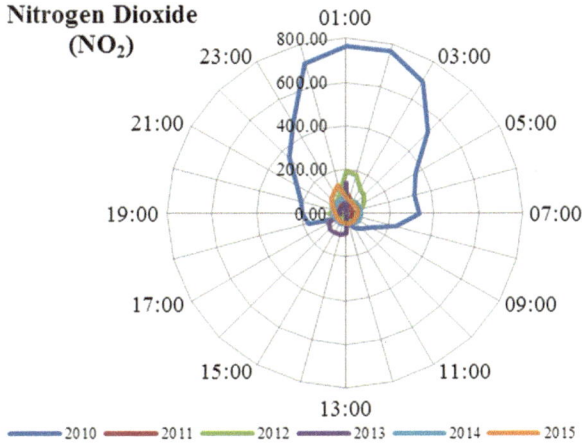

Fig. 7 Diurnal variation in concentration of nitrogen dioxide (NO₂)

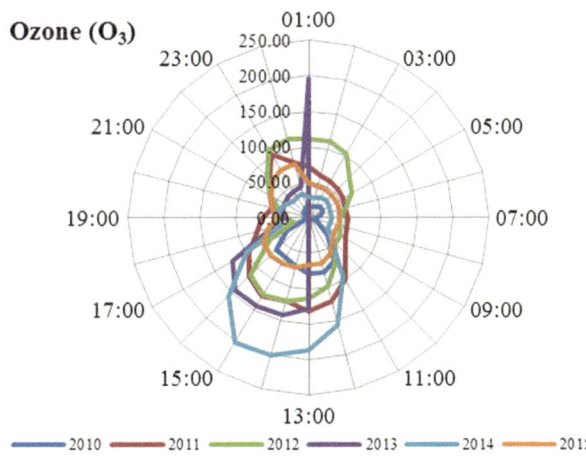

Fig. 8 Diurnal variation in concentration of ozone (O₃)

bursting firecrackers as considerable amount of light is produced by firecrackers which results in considerable night time surface ozone formation. Nishanth et al. (2012) and Peshin et al. (2017) also reported high concentration of PM, NO_2, and also O_3 during nighttime on Diwali eve. This shows that the nighttime celebration during Diwali day and bursting of huge quantity of firecrackers results in high pollutant concentration during nighttime; moreover, as boundary layer height is low during night this further aggravates the concentration and piling of pollutants.

3.3 Post-Diwali Dispersion of Pollutants

A forward Trajectory Hysplit model was computed to study the pattern of dispersion of pollutants after Diwali. For this, trajectories at different altitude (10 m, 100 m, and 1000 m) were studied 7 days succeeding the Diwali date in Delhi from

2010 to 2015. The 6 year trajectory shows different pattern of dispersions. Pollutants are dispersing off towards western part of India entering Arabian Sea in year 2010, 2013, and 2016. In all these years Diwali was in late October or in early November. In years 2012 and 2015 when Diwali was in mid-November most of the wind dispersion were moving from eastern to south eastern and then turning towards western part. Interestingly in year 2011 and 2014 a restricted dispersion was observed at 10 m and 100 m height and most of the winds were reverting back towards Delhi thus concentrating pollution in and near Delhi region. The Diwali was celebrated around 20th October during these 2years.

A report by Weather Spark (weatherspark.com n.d.) states that the lowest wind speed or calm environment was generally reported around October 16–20, which inhibits proper dispersion of pollutants. Also, a study reported by Guttikunda and Gurjar (2012), on the effect of meteorology on urban pollution dispersion depicted that wind speed in mid-October decreases a little (0.2 m/s, month to month) and then increases (0.3 m/s, month to month) by early November. The Hysplit study shows that the pollution released during the Diwali days is either dispersed towards western or eastern part of India depending on the part of month in which it is celebrated. Especially if the festival is celebrated in mid-October the chances of restriction of pollution near Delhi increases which can be studied further. However, local meteorology plays a crucial role in deciding the pollution level of an area during and after Diwali celebration (Fig. 9).

3.4   Air Quality Health Index (AQHI)

Three-hour moving maximum average was computed for criteria pollutants, particulate matter (PM), NO_2, and O_3 for the eve of Diwali for 6 years, 2010–2015. After indexing the multiple pollutants, year 2010 was found to be falling in Very High (10^+) category of the AQHI, followed by 2013 and 2014 in High Health risk category (8). AQHI categories between 7 and 10^+, viz., High and Very High Health risk, follow a health remark to take precautionary actions to avoid the outdoor air exposure by rescheduling or cancelling outdoor exposure events even for general population.

Years 2011 and 2015 indexed in moderate health risk category (5), and year 2013 had low health risk (3) on the day of Diwali. AQHI categories, with values between 4 and 6, viz., Moderate Health risk, communicate precautionary actions for sensitive groups of the population (elderly, children below age 5, and respiratory and cardiovascular patients) like avoiding outdoor activities if coughing and throat irritation occurs.

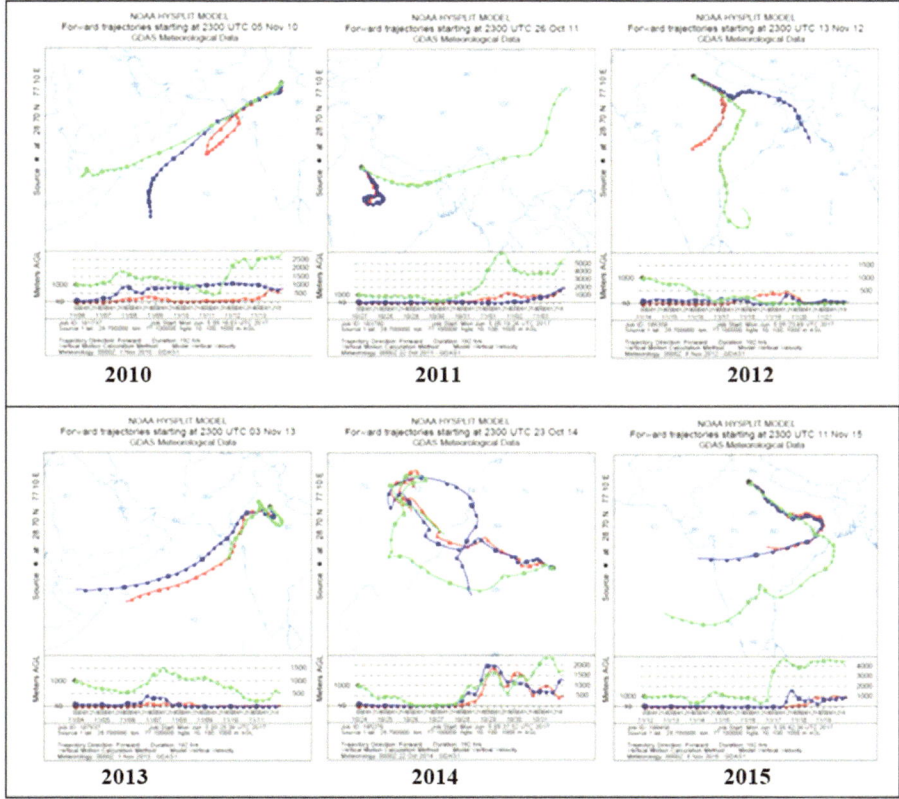

Fig. 9 Hysplit model, Backward trajectories

4 Conclusion

Exhibition of fireworks during Diwali celebration causes massive though short-lived air pollution. Concentration of pollutants showed acute rise, especially of particulate matter whose concentration increased many times as we start to burst crackers in the evening of Diwali. The short-term exposure of these pollutants beyond the permissible limits can increase the probability of acute health effects, especially for the vulnerable group of population. Therefore, it is of utmost importance to monitor the outburst of pollution by such events. Although regulating or restricting the festival would be socially unacceptable, adapting to cleaner, more environmental friendly manner to celebrate the festival is the need of the hour.

Acknowledgement The authors deeply acknowledge the Rajiv Gandhi National Fellowship for Disabled (RGNFD), SAFAR programme and Ministry of Earth Sciences (Project Order No. MOES/16/07/2013-RDEAS) for providing academic and financial support for the study.

References

Abelsohn A, Stieb DM (2011) Health effects of outdoor air pollution, approach to counseling patients using the air quality health index. Can Fam Physician 57(8):881–887

Ali K, Momin GA, Tiwari S, Safai PD, Chate DM, Rao PSP (2004) Fog and precipitation chemistry at Delhi, North India. Atmos Environ 38:4215–4222

Attri AK, Kumar U, Jain VK (2001) Microclimate: formation of ozone by fireworks. Nature 411(6841):1015

Committees of the Environmental and Occupational Health Assembly (COEHA) of the American Thoracic Society (1996) Am J Respir Crit Care Med:131–133

Dockery DW, Pope CA, Xu X, Spengler JD, Ware JH, Fay ME, Ferris BG Jr, Speizer FE (1993) An association between air pollution and mortality in six U.S. Cities. N Engl J Med 329:1753–1759

Guttikunda SK, Gurjar BR (2012) Role of meteorology in seasonality of air pollution in megacity Delhi, India. Environ Monit Assess 184(5):3199–3211

Hoet PHM, Brüske-Hohlfeld I, Salata OV (2004) Nanoparticles—known and unknown health risks. J Nanobiotechnology 2:12

Huang SL, Hsu MK, Chan CC (2003) Effects of submicrometer particle compositions on cytokine production and lipid peroxidation of human bronchial epithelial cells. Environ Health Prespect 111:478–482

Kumar M, Singh RK, Murari V, Singh AK, Singh RS, Banerjee T (2016) Fireworks induced particle pollution: a spatio-temporal analysis. Atmos Res 180:78–91

Mandal P, Prakash M, Bassin JK (2012) Impact of Diwali celebrations on urban air and noise quality in Delhi City, India. Environ Monit Assess 184:209–215

Nishanth T, Praseed KM, Rathnakaran K, Satheesh Kumar MK, Krishna RR, Valsaraj KT (2012) Atmospheric pollution in a semi-urban, coastal region in India following festival seasons. Atmos Environ 47:295–306

Pal R, Mahima, Gupta A, Singh C, Tripathi A, Singh RB (2013) The effects of fireworks on ambient air and possible impact on cardiac health, during Deepawali festival in North India. World Heart J 5(1):21–32

Parkhi N, Chate D, Ghude SD, Peshin S, Mahajan A, Srinivas R, Surendran D, Ali K, Singh S, Trimbakh H, Beig G (2016) Large inter annual variation in air quality during the annual festival 'Diwali' in an indian megacity. J Environ Sci 43:265–272

Peshin SK, Sinha P, Bisht A (2017) Impact of Diwali firework emissions on air quality of New Delhi, India during 2013–2015. Mausam 68(1):111–118

Pope CA 3rd, Schwartz J, Ransom MR (1992) Daily mortality and PM10 pollution in Utah Valley. Arch Environ Health 47(3):211–217

Ravindra K, Mittal AK, Van Grieken R (2001) Health risk assessment of urban suspended particulate matter with special reference to polycyclic aromatic hydrocarbons: a review. Rev Environ Health 16(3):169–189

Ravindra K, Mor S, Kaushik CP (2003) Short-term variation in air quality associated with firework events: a case study. J Environ Monit 5:260–264

Sarkar S, Khillare PS, Jyethi DS, Hasan A, Parween M (2010) Chemical speciation of respirable suspended particulate matter during a major firework festival in India. J Hazard Mater 184:321–330

Schwartz J (1993) Air pollution and daily mortality in Birmingham, Alabama. Am J Epidemiol 137(10):1136–1147

Singh DP, Gadi R, Mandal TK, Dixit CK, Singh K, Saud T et al (2010) Study of temporal variation in ambient air quality during Diwali festival in India. Environ Monit Assess 169(1):1–13

Stein AF, Draxler RR, Rolph GD, Stunder BJB, Cohen MD, Ngan F (2015) NOAA'S hysplit atmospheric transport and dispersion modeling system. Am Meteorol Soc:2059–2078

Stieb DM, Doiron MS, Blagden P, Burnett RT (2005) Estimating the public health burden attributable to air pollution: an illustration using the development of an alternative air quality index. J Toxicol Environ Health 68:1275–1288

Stieb DM, Burnett RT, Smith-Doiron M, Brion O, Shin HH, Economou V (2008) A new multipollutant, no-threshold air quality health index based on short-term associations observed in daily TimeSeries analyses. J Air Waste Manage Assoc 58:435–450

Thakur A, Bhatia RK (2015) Study of urban air quality—a case study of Delhi. Int J Sci Eng Technol 3(5):1151–1153

Tiwari S, Chate DM, Srivastava MK, Safai PD, Srivastava AK, Bisht DS, Padmanabhamurty B (2012) Statistical evaluation of PM10 and distribution of PM1, PM2.5, and PM10 in ambient air due to extreme fireworks episodes (Deepawali festivals) in megacity Delhi. Nat Hazards 61:521–531

To T, Shen S, Atenafu EG, Guan J, McLimont S, Stocks B et al (2013) The air quality health index and asthma morbidity: a population-based study. Environ Health Perspect 121:46–52

Vernath JM, Reilly CA, Vernath MM, Moss TA, Langelier CR, Lanza DL, Yost GS (2004) Inflammatory cytokines and cell death in BEAS-2B lung cells treated with soil dust, lipopolysaccharide, and surface modified particles. Toxicol Sci 82:88–96

Weatherspark.com—Average Weather in New Delhi India (Report) (n.d.). (https://weatherspark.com/m/109174/10/Average-Weather-in-October-in-New-Delhi-India) Date of accession—June

WHO (2014) http://www.who.int/phe/health_topics/outdoorair/databases/cities-2014/en/

Mushrooms as a Biological Tool in Mycoremediation of Polluted Soils

Monika Thakur

Abstract One of the major environmental problems faced by today's world is the contamination of soil, water, and air by toxic chemicals, and the distinct and unique role of microorganisms in the detoxification of polluted soil and environments is well recognized. Fungal mycelia have been primary governors for maintaining ecological equilibrium because they control the flow of nutrients. The strength and health of any ecosystem is a direct measure of its main components—the fungal populations and their interaction with other organisms such as plants, animals, and bacteria. Using fungi as the starter culture species in a mycoremediation project sets the stage for other organisms to participate in the rehabilitation process. The introduction of fungal mycelium into a polluted site triggers a flow of activity and begins to replenish the polluted ecosystem. Mycoremediation is an economically and environmentally sound alternative for bioremediation. It is not widely used at present, but this technology has wider potential than other technologies. Fungi perform a wide variety of functions in ecosystem and potentially have been proven to be clean, simple, and relatively inexpensive for environmental remediation. Examples of fungi used as mycoremediators are *Pleurotus ostreatus; Rhizopus arrhizus; Phanerochaete chrysosporium and P. sordida;* and *Tramates hirsuta and T. versicolor; and Lentinus edodes and L. tigrinus. Thus, this clean technology has greater potential and its untapped potential has to be fully exploited.*

Keyword Bioremediation · Mycoremediation · In situ · Ex situ · Mushroom mycelium · Polluted soil · Underexploited

M. Thakur (✉)
Amity Institute of Food Technology, Amity University,
Noida, Uttar Pradesh, India
e-mail: mthakur1@amity.edu

© The Author(s), under exclusive licence to Springer Nature Switzerland AG 2019   27
T. Jindal, *Emerging Issues in Ecology and Environmental Science*,
SpringerBriefs in Environmental Science, https://doi.org/10.1007/978-3-319-99398-0_3

1 Introduction

One of the major environmental problems faced by today's world is the contamination of soil, water, and air by various toxic chemicals as a result of industrialization and extensive use of chemicals in agriculture. There are various "**clean technologies**" which emphasize on reduced waste generation, treatment and conversion of waste into some useful form. These clean technologies emphasize on the use of various biological methods for the remediation of waste. The soil is getting more and more polluted day by day. Remediation of the polluted soils is a challenging job. **Bioremediation** is a treatment process that uses naturally occurring microorganisms to break down, or degrade, hazardous substances into less toxic or nontoxic substances. Bioremediation is an attractive technology that utilizes the metabolic potential of microorganisms in order to clean up the environmental pollutants to the less hazardous forms with less input of chemicals, energy, and time (Asgher et al. 2008; Haritash and Kaushik 2009).

The microbes used to perform the function of bioremediation are known as **bioremediators**. "To bioremediate" means to use living things to reclaim contaminated environments. The introduction of exogenous microorganisms into environments— **bioaugmentation**, has been used as an attempt to accelerate bioremediation (Watanabe 2001). Some microorganisms that live in soil and groundwater naturally eat certain toxic chemicals that are harmful to the environment. Watanabe (2001) reported that many naturally occurring microbes have been utilized in a variety of bioremediation processes.

One such biological clean technology is "mycoremediation" which is based on the use of fungi for the removal of waste and toxins from the environment. **Mycoremediation** is one of the most successful and technical areas of bioremediation, which refers specifically to the use of fungal mycelium (Singh 2006). It is a process of using fungi to degrade the contaminants in the polluted environment. Jagtap et al. (2003) discussed mycoremediation as a bioremediation process of using fungi (saprophytic, parasitic, and mycorrhizal) to remove pollutants from the environment. Fungal mycelium stimulates microbial and enzyme activity, and thus reduces in situ production of toxins. The potential applications for mycoremediation technologies have been reported from time to time. Fungal species have been shown to accumulate toxic metals, and even rare earth elements. Fungi are great biodegrades and the resultant compost has been used to enhance the growth of plants as well as bioremediation activity in the environment (Jagtap et al. 2003). Mycelia of fungi are unique among microorganism having the ability to enhance plant growth. They secrete variety of extracellular enzymes involved in pollutants degradation. Some fungi are hyperaccumulators, and are capable of absorbing and concentrating heavy metals in the fruiting bodies of mushrooms (Jagtap et al. 2003). Mycoremediation process involve mixing of mycelium into contaminated soil, placing mycelial mats over toxic sites, or and even the combination of these two techniques. Mycoremediation has been applied to oil spills, contaminated and polluted soil, industrial chemicals, contaminated water, and even farm wastes (Bennet et al.

2001). Bioremediation technology leads to degradation of pollutants and may be a lucrative and environmentally beneficial alternative (Thakur 2015).

Macrofungi (mushrooms) and other microfungi possess enzymes for the degradation of variety of pollutants (Purnomo et al. 2013; Kulshreshtha et al. 2013). However, mushrooms are becoming more popular nowadays for bioremediation process because they are not only a good bioremediation tool but also provide nutritional and health benefits (Kuforiji and Fasidi 2008; Zhu et al. 2013; Thakur 2014; Thakur 2015). Their multidirectional role has now attracted researchers to work in the field of mushroom cultivation and mycoremediation.

In the present scenario, the pollution is increasing at a faster pace. The content of toxic and heavy metals in the environment is increasing day by day and is emerging as a serious problem. In the past, various other treatment methods like thermal, chemical, and physical have failed to reduce or eliminate the pollution problem because those methods can only shift the pollution to a new phase, i.e., air pollution and make air more polluted (Williams et al. 1992). The simultaneous cleanup of all these contaminants by thermal/chemical/conventional method is technically difficult and expensive, and these very methods can also destroy other soil biotic components (Dua et al. 2006). Therefore, mycoremediation is an emerging cleanup technology for polluted sites. Keeping this in mind, the chapter discuss the use of mushroom species as a biological tool for cleanup and degradation of waste and pollutants present in the environment. This chapter aims to study the role of macrofungi and exploit their potential to remediate the polluted soil.

2 Fungi Role in Mycoremediation

Fungi have been used in many diverse applications since ancient times. They have been the major decomposers of various complex polymers as cellulose, hemicellulose, and lignin in the ecosystem (D'Annibale et al. 2006). Macrofungi also plays an important role as natural environment remediator (Pletsch et al. 1999; Matsubara et al. 2004; Thakur 2015). Mycoremediation is an innovative biotechnological application that uses fungus for in situ and ex situ cleanup and management of contaminated sites (Thomas et al. 2009). Mushroom has been used for nutritional and medicinal purposes since times immemorial. Fungi have been nature's most powerful decomposers, secreting strong enzymes. They have also known as mycoremediation tool because of their use in bioremediation of different types of pollutants. They have fast mycelial growth, great biomass production and extensive hyphae which will grow and reach deeper components of environment. This clean technology is based on efficient enzymes (cellulases, hemicellulases, lignin-degrading enzymes, etc.) produced by various mushroom species. These fungi have also been known to transform a wide variety of hazardous pollutants by the process of bioremediation (Alexander 1994; Ashoka et al. 2002).

Fungal mycelia have been primary governors for maintaining ecological equilibrium because they control the flow of nutrients. The strength and health of any ecosystem is dependent upon basic thing—the fungal populations and their interaction with living organisms (producers, consumers, and decomposers). The introduction of fungal mycelium as "starter culture" into a polluted site triggers a flow of enzymes and begins to replenish the polluted ecosystem.

Mycoremediation is an economically and environmentally sound alternative for bioremediation. It restores depleted and polluted soils. Currently, burning, hauling, and constructing new buildings have been the common practices to remove or clean up toxic wastes. But because of these processes, environments do not get rid of all the waste materials and slowly but steadily the pollutants have been added to back to the soil only. This leaves the soil lifeless and contaminated. Various toxins (including mercury, PCBs, and dioxins) are added to our food chain, and become more concentrated at each and every step. Fungal mycelia can destroy these toxins in the soil before they enter our food supply chain. Mycoremediation is thus a biological mechanism to destroy, transform, or immobilize environmental contaminants (Adenipekun and Lawal 2012).

Mycoremediation is not widely used at present, but this technology has wider potential. Fungi perform a wide variety of functions in ecosystem and potentially prove to be clean, simple, and relatively inexpensive for environmental remediation (Kulshrestha et al. 2014). Loske et al. (1990) reported the main contaminants of polluted soils mainly polycyclic aromatic hydrocarbons (PAH's); polychlorinated biphenyls (PCB's), and dioxins.

3 Process of Mycoremediation

Mushrooms use different methods to decontaminate polluted sites and stimulate the environment, such as biodegradation, biosorption, and bioconversion.

3.1 Biodegradation

The biodegradation mechanism is very complex process. In this process, there is degradation and recycling of complex molecules to its mineral constituents. The term "*Biodegradation*" is used to describe the ultimate degradation and recycling of complex molecule to its simpler mineral constituents. It is the process which leads to complete mineralization of the complex compound to simpler ones like CO_2, H_2O, NO_3, and other inorganic compounds by living organisms. Table 1 enlists the enzymes secreted and degradation abilities of various mushroom species. Mushrooms produce various extracellular enzymes such as peroxidases, ligninase (lignin peroxidase, manganese-dependent peroxidase, and laccase), cellulases, pectinases, xylanases, and oxidases. These enzymes are induced by their substrates and are capable of degrading nonpolymeric, recalcitrant pollutants such as

Table 1 Role of mushroom species in biodegradation of pollutants

Sr. No.	Mushroom Species	Role as a mycoremediator	References
1	*Pleurotus ostreatus*	Oxo-biodegradable plastic: mushroom species growing on plastic degraded the plastic.	da Luz et al. (2013)
2	*Lentinula edodes*	Mushroom species degraded 2,4-dichlorophenol (DCP) by using vanillin as an activator.	Tsujiyama et al. (2013)
3	*Pleurotus pulmonarius*	Radioactive cellulosic-based waste: Waste-containing mushroom mycelium was solidified with cement and then the solidified waste acted as the first barrier against the release of radio-contaminants from the site.	Eskander et al. (2012)
4	*Auricularia sp.*, *Schizophyllum commune*, and *Polyporus sp.*	Malachite green dye was degraded by the mushroom species in 10 days.	Rajput et al. (2011)
5	*Pleurotus pulmonarius*	This mushroom species helps in the degradation of crude oil.	Olusola and Anslem (2010)
6	*Coriolus versicolor*	This mushroom species possesses the ability to degrade PAH with the help of lignin-modifying enzymes laccase, manganese-dependent peroxidase (MnP), and lignin peroxidase (LiP).	Jang et al. (2009)

nitrotoluenes, PAHs, organic and synthetic dyes, pentachlorophenol (VanAcken et al. 1999; Hammel et al. 1991; Johannes et al. 1996; Ollikka et al. 1993; Heinfling et al. 1998; Lin et al. 1990; Haritash and Kaushik 2009) under in vitro conditions. Recently, it has also been reported that various mushroom species are also able to degrade polymers such as plastics (da Luz et al. 2013).

3.2 Biosorption

Biosorption is the process for the removal for the removal of pollutants from the environment with the help of mushroom. It has been considered as an alternative to the remediation of industrial effluents as well as the recovery of metals present in effluent. Biosorption is a process based on the sorption of metallic ions/pollutants/ xenobiotics from effluents by live or dried biomass which often exhibits a marked tolerance towards metals and other adverse conditions.

Several chemical processes may be involved in biosorption like adsorption, ion exchange processes, and covalent binding. The polar groups of proteins, amino acids, lipids, and structural polysaccharides (chitin, chitosan, glucans) may be involved in the process of biosorption. Table 2 enlists the biosorptive capacity of biomass of mushroom species. They also reported that the biosorption capacity of dead biomass is greater/similar to/less than that of living cells.

Table 2 Removal of pollutants by biomass of mushroom using biosorption process

S. no	Mushroom species	Pollutants and role of mushroom species	References
1	*Agaricus bisporus,* *Lactarius piperatus*	Cadmium (II) ions: *mushroom species* showed higher removal efficiency on Cd(II) ions	Nagy et al. (2013)
2	*Fomes fasciatus*	Copper (II): Mushroom is efficient in biosorption of Cu (II) ions	Sutherland and Venkobachar (2013)
3	*Pleurotus platypus,* *Agaricus bisporus,* *Calocybe indica*	Copper, zinc, iron, cadmium, lead, nickel: Mushroom species are efficient biosorbents for the removal these ions from aqueous wastes.	Lamrood and Ralegankar (2013)
4	*Flammulina velutipes*	Copper: Mushroom fruiting body used as biosorbents for removing copper ions from aqueous wastes.	Luo et al. (2013)
5	*Pleurotus tuber-regium*	Heavy metals: *Mushroom species* bioabsorbed pollutants (heavy metals) from soils artificially contaminated with some heavy metals.	Oyetayo et al. (2012)
6	*Pleurotus ostreatus*	Cadmium: Mushroom species bioabsorbed cadmium ions from the substrate.	Tay et al. (2011)
7	*Pleurotus sajor-caju*	Mushrooms bioabsorbed heavy metals.	Jibran and MilseeMol (2011)

3.3 Bioconversion

In this process there has been conversion of industrial waste into some mushroom species. The lignocellulosic waste, generated by industries, can be used for cultivation of mushroom which can be further used as a product. Mushroom species cultivated on industrial and agroindustrial wastes are given in Table 3 (Kulshreshtha et al. 2010; Kulshreshtha et al. 2013).

4 Potential of Mushrooms in Mycoremediation

Although bioremediation by bacterial agents has received attention of many researchers, the role of fungi has been still inadequately explored. The ability of fungi to transform a wide variety of hazardous chemicals has aroused interest in using them for bioremediation. Mushroom forming fungi are amongst nature's most powerful decomposers, secreting strong extra cellular enzymes due to their mycelial growth and biomass production (Elekes and Busuioc 2010). These enzymes include lignin peroxidases (LiP), manganese peroxidase (MnP), and laccase. Thus, carbon sources such as sawdust, straw, and corncob can be used to enhance degradation rates by these organisms at various polluted sites (Adenipekun and Lawal 2012). *Phanerochaete chrysosporium, Agaricus bisporus, Trametes versicolor, Pleurotus*

Table 3 Bioconversion of waste by mushroom species

S. No.	Waste material	Mushroom species	Mushroom cultivation	References
1	Handmade paper, cardboard, and industrial waste	*Pleurotus citrinopileatus*	Successfully cultivated. Basidiocarps possessed good nutrient content and no genotoxicity	Kulshreshtha et al. (2013)
2	Sawdust of different woods	*Pleurotus ostreatus*	Biomass of mushroom has been produced in submerged liquid culture were analyzed	Akinyele et al. (2012)
3	Agroindustrial residues such as cassava, sugar beet pulp, wheat bran, and apple and pear pomace	*Volvariella volvacea*	Enzyme activities were measured during the fermentation of substrates	Akinyele et al. (2011)
4	Handmade paper, cardboard and industrial waste	*Pleurotus Florida*	Successfully cultivated. Basidiocarps possessed normal morphology and no genotoxicity	Kulshreshtha et al. (2010)
5	Cotton waste, rice straw, cocoyam peels and sawdusts of *Mansonia altissima, Boscia angustifolia*, and *Khaya ivorensis*	*Pleurotus*	Successfully cultivated with good crude protein, fat, and carbohydrate contents in fruiting bodies.	Kuforiji and Fasidi (2009)
6	Paddy straw, sorghum stalk, and banana pseudostem	*Pleurotus eous* and *Lentinus connatus*	Waste successfully bio-converted by mushroom with good biological efficiency	Rani et al. (2008)
7	*Terminalia superba, Mansonia altissima, Holoptelea grandis,* and *Miliciaex excelsa*	*Pleurotus tuber-regium*	Mushroom species grown on trees	Jonathan et al. (2008)
8	Cotton waste, sawdust of *Khaya ivorensis* and rice straw	*Pleurotus tuber-regium*	Sclerotia propagated on groundnut shells and cocoyam peels with lipase and phenoloxidase; cellulase, carboxy-methylcellulase enzymatic activities	Kuforiji and Fasidi (2008)
9	Eucalyptus waste	*Lentinula edodes*	Successfully converted this waste and qualitative and quantitative changes were also measured	Brienzo et al. (2007)
10	Mushroom fruiting bodies were grown on vineyard prunings, barley straw and wheat straw	*Lentinula edodes*	Bioconversion of waste having highest biological efficiency, yield, and shortest production cycle	Gaitán-Hernández et al. (2006)

(continued)

Table 3 (continued)

S. No.	Waste material	Mushroom species	Mushroom cultivation	References
11	Wheat straw	*Lentinula tigrinus*	Characterized the production of lingocellulosic enzymes and bioconverted the wheat straw into fruiting bodies	Lechner and Papinutti (2006)
12	Banana leaves	*V. volvacea*	Efficient bioconversion with good yield of fruiting bodies	Belewu and Belewu (2005)

ostreatus, and other many mushroom species have been reported for decontamination of polluted sites (Adenipekun and Lawal 2012; Thakur 2015). Mushrooms have long been known for their nutritive and medicinal benefits. Sesli and Tuzen (1999) reported that mushrooms can be used to evaluate the level of environmental pollution and to remediate metal-polluted soils. Also, many studies have been carried out to evaluate the possible threats to human health from the ingestion of mushrooms containing heavy metals (Tismal et al. 2010; Ouzouni et al. 2009; Sesli and Tuzen 1999).

Based on literature and research, white rot fungus accounts for almost 30% of the total research on fungi used in bioremediation process (Adenipekun and Lawal 2012). White rot fungi have been used for bioremediation of pesticides, degradation of petroleum hydrocarbons and lignocellulolytic wastes in the pulp and paper industry. White rot fungi are excellent mycoremediators of toxins held together by hydrogen–carbon bonds. Enzymes secreted by white rotters include lignin peroxidases, manganese peroxidases, and laccases.

Some specific examples of macrofungi mycelium especially white rot fungus used for mycoremediation are (Fig. 1):

Sr. No.	Macrofungi as mycoremediator	References
1	*Phanerochaete chrysosporium*	Leonardi et al. (2007), Adenipekun and Lawal (2012), Sasek and Cajthaml (2005), Nigam et al. (1995), Aitken and Irvine (1989), Bumpus et al. (1985), Barr and Aust (1994)
2	*Lentinus edodes*	Adenipekun and Lawal (2012)
3	*Lentinus tigrinus*	Stella et al. (2012)
4	*Lentinus squarrosulus* (Mont.) Singer	Adenipekun and Fasidi (2005) and Adenipekun and Isikhuemhen (2008)
5	*Pleurotus ostreatus* (Jacq. Fr.) P. Kumm	Sack and Gunther (1993), Bojan et al. (1999), Sykes (2002), Okparanma et al. (2011), Baldrian et al. (2000), Eggen and Majcherczyk (1998), Eggen and Sveum (1999), Bhattacharya et al. (2012)
6	*Pleurotus tuber-regium* (Fries) Singer	Isikhuemhen et al. (2003), Adenipekun et al. (2011a)
7	*Pleurotus pulmonarius*	Adenipekun et al. (2011b)
8	*Trametes versicolor*	Stamets (2010), Tanaka et al. (1999), Novotny et al. (2004), Morgan et al. (1991), Gadd (2001)
9	*Bjerkandera adusta*	Adenipekun and Lawal (2012), Pozdnyakova (2012)
10	*Irpex lacteus*	Bhatt et al. (2002), Adenipekun and Lawal (2012)

Fig.1 (**a**) *Phanerochaete chrysosporium,* (**b**) *Lentinus edodes,* (**c**) *Lentinus tigrinus,* (**d**) *Lentinus squarrosulus* (Mont.) Singer, (**e**) *Pleurotus ostreatus* (Jacq. Fr.) P. Kumm; (**f**) *Pleurotus tuber-regium* (Fries) Singer (**g**) *Pleurotus pulmonarius;* (**h**) *Trametes versicolor;* (**i**) *Bjerkandera adusta;*(**j**) *Irpex lacteus*

5   Advantages of Mycoremediation

Mycoremediation technologies help in fungal species growth and increase its population by creating optimum environmental conditions for them to detoxify the maximum amount of contaminants. A fungus produces various nonspecific enzymes which can act on wide variety of environment pollutants. Hyphae allow fungi to expand their surface area, make them easier to contact the pollutant. There have been numerous advantages of using mycoremediation over commercialized technologies, including the following:

*   Public acceptance.
*   Natural and environment friendly.
*   Safety.
*   Simple and quiet.
*   Low maintenance.
*   Reusable end products.
*   Low cost.
*   Flexibility.
*   Fast.

6   Constraints for Mycoremediation

The use of macrofungi like mushrooms for remediation of polluted soils has been known in the recent years. This mycoremediation is not only a clean technology but also generate fruiting bodies. Thus the mushroom production can generate not only a healthy food but also the food for livelihood. Research has shown that mushroom species like *P. ostreatus* and *P. chrysosoporium* have emerged as model systems for studying bioremediation. But, a great deal still remains to be learned about the basic knowledge of how this white-rot fungus removes pollutants. The majority of mycoremediation work has been done on *Phanerochaete chrysosporium*. The fungus has the potential of mycoremediation because of its lignin-degrading enzyme system. A similar degrading ability has been described by other species of white rot fungus but not with the degree of success reported for *Phanerochaete chrysosporium*. Sasek (2003) reported that the performance of white rot fungus in soil bioremediation depends upon its survival in the soil environment, colonization, relationship and interaction with other soil microflora. Still, research trials on other fungi are unexplored and underexploited.

Mycoremediation is a very important process but still there are various problems that are hindering its widespread use. Boopathy (2000)) discussed some of the factors limiting bioremediation technologies. Various challenges faced are:

*   Contamination by other fungi (*Penicillium* spp., *Aspergillus* spp.) while interaction with them.

- Fungal species are unable to compete with other native microbes in soils. Some bacteria could either inhibit the growth of fungi, or in combination with fungi enhance degradation of pollutants.
- Nutrient cycle should be completely understood.
- Starter cultures: The problem to be borne in mind is that in bioremediation projects mushroom mycelium should not be used as a starter material.
- Legal issues: There are also legal issues in this process. There are several patents specifically granted for matching fungus against a toxin. This is a major hindrance in preventing wide-scale fungal cleanup of toxins from polluted sites.
- Mushroom cultivation process: The lack of experienced mushroom cultivators in outdoor trials is a problem in mycoremediation. This lacking has affected the success of several trials.

Therefore, though there are many constraints in this effective technology, it has proven to be a boon to the soil. More research is still required for further exploration.

7   Future Prospects

In recent advancements the addition of required fungal strains to the soil, the enhancement of the indigenous microbial population and its ability to break down various exposures contaminants have proven successful. Whether the fungal mycelia are native or newly introduced to the site, the process of destroying contaminants is important and critical for understanding mycoremediation. There is no definite time frame for complete mycoremediation as the time taken by various contaminants and types of applications will vary. That is why the research in this area is still in the experimental phase and unexplored. Further, the application of this technology in large scale projects will demand much more work to streamline the methodologies.

Once the research and technology is complete, it will have a wider application. With appropriate funding by the government/semi-government and other organizations, the technology could be developed and made available for commercialization. However, current funding has been limited. But extensive research needs to be pursued as the technology has proven successful in micro-sites. This cleaner technology has been expected to be faster and more cost-effective than other remediation technologies once it is commercialized. The use of fungi for remediation would allow commercial concern to offer inexpensive, safe products to their customers. If the underexploited potential of fungus mycelium is further exploited, it will go a long way in eradicating pollution from soils. Thus the mission of the pollution-free environment can be achieved and our future generations will have a better environment to live.

8   Conclusion

Fungi can be used an effective tool to reduce waste materials in contaminated soils via nonspecific enzymes activities. Evidences have shown that mushrooms have the potential to clean up soils contaminated with various toxic elements. Mycoremediation is not a panacea, but an effective and powerful tool to remediate soil pollution. Mushrooms have tremendous potential to be used in bioremediation process. The cultivation of edible mushroom species on agricultural and industrial wastes may thus be a value-added process capable of converting these wastes into foods (mushrooms). Besides producing nutritious mushrooms, it reduces genotoxicity and toxicity of contaminated sites. Mycoremediation through mushroom cultivation will alleviate two of the world's major problems, i.e., waste accumulation and production of proteinaceous food, simultaneously. Thus, there is a need for further research towards the exploitation of potential of the mushroom species as a bioremediation tool and its safety aspects for consumption as a food product.

References

Adenipekun CO, Fasidi IO (2005) Bioremediation of oil polluted soil by *Lentinus subnudus*, a Nigerian white rot fungus. Afr J Biotechnol 4(8):796–798

Adenipekun CO, Isikhuemhen OS (2008) Bioremediation of engine oil polluted soil by the tropical white-rot fungus, Lentinussquarrosulus Mont (Singer). Pak J Biol Sci 11(12):1634–1637

Adenipekun CO, Lawal R (2012) Uses of mushrooms in bioremediation: a review. Biotechnol Mol Biol Rev 7(3):62–68

Adenipekun CO, Ejoh EO, Ogunjobi AA (2011a) Bioremediation of cutting fluids contaminated soil by *Pleurotus tuber-regium* singer. Environmentalist 32:11–18

Adenipekun CO, Ogunjobi AA, Ogunseye OA (2011b) Management of polluted soils by a white-rot fungus, *Pleurotus pulmonarius*. Assump Univ J Technol 15(1):57–61

Aitken MB, Irvine RL (1989) Stability testing of ligninase and Mn peroxidase from *Phanerochaetechrysosporium*. Biotechnol Bioeng J 34:1251–1260

Akinyele BJ, Olaniyi OO, Arotupin DJ (2011) Bioconversion of selected agricultural wastes and associated enzymes by *Volvariella volvacea*: an edible mushroom. Res J Microbiol 4:63–70

Akinyele JB, Fakoya S, Adetuyi CF (2012) Anti-growth factors associated with *Pleurotus ostreatus* in a submerged liquid *fermentation*. Malays J Microbiol 4:135–140

Alexander M (1994) Biodegradation and bioremediation, 2nd edn. Academic Press, San Diego

Asgher M, Bhatti HN, Ashraf M, Legge RL (2008) Recent developments in biodegradation of industrial pollutants by white-rot fungi and their enzyme system. Biodegradation 19:771–783

Ashoka G, Geetha MS, Sullia SB (2002) Bioleaching of composite textile dye effluent using bacterial consortia. Asian J Microbial Biotechnol Environ Sci 4:65–68

Baldrian P, Der Wiesche CI, Gabriel J, Nerud F, Zadrazil F (2000) Influence of cadmium and mercury on activities of ligninolytic enzymes and degradation of polycyclic aromatic hydrocarbons by *Pleurotus ostreatus* in soil. Appl Environ Microbiol 66:2471–2478

Barr BP, Aust D (1994) Mechanisms of white-rot fungi use to degrade pollutant. Environ Sci Technol 28:78–87

Belewu MA, Belewu KY (2005) Cultivation of mushroom (*Volvariella volvacea*) on banana leaves. Afr J Biotechnol 4:1401–1403

Bennet JW, Connick WJ, Daigle D, Wunch K (2001) Formulation of fungi for in situ bioreme- diation. In: Gadd GM (ed) Fungi in bioremediation. Cambridge University Press, Cambridge, pp 97–108

Bhatt M, Cajthaml T, Sasek V (2002) Mycoremediation of PAH-contaminated soils. Folia Microbiol 47(3):255–258

Bhattacharya S, Angayarkanni J, Das A, Palaniswamy M (2012) Mycoremediation of benzo[a] pyrene by Pleurotus ostreatus isolated from Wayanad District in Kerala, India. Int J Pharm Bio Sci 2(2):84–93

Bojan BW, Lamar RT, Burjus WD, Tien M (1999) Extent of humification of anthrecene, fluoran- thene adbenzo (a) pyrene by Pleurotus ostreatus during growth in PAH-contaminated soils. Lett Appl Microbiol 28:250–254

Boopathy R (2000) Factors limiting bioremediation technologies. Bioresour Technol 74(1):63–67

Brienzo M, Silva EM, Milagres AM (2007) Degradation of eucalyptus waste components by Lentinula edodes strains detected by chemical and near-infrared spectroscopy methods. Appl J Biochem Biotechnol 4:37–50

Bumpus JA, Tien M, Wright D, Aust SD (1985) Oxidation of persistent environmental pollutants by a white rot fungus. Science 228:1434–1436

D'Annibale A, Rosetto F, Leonardi V, Federici F, Petruccioli M (2006) Role of autochthonous filamentous Fungi in bioremediation of a soil historically contaminated with aromatic hydro- carbons. Am Soc Microbiol 72(1):28–36

Da Luz JMR, Paes SA, Nunes MD, da Silva MCS, Kasuya MCM (2013) Degradation of Oxo- biodegradable plastic by Pleurotus ostreatus. PLoS One 4(8):69386

Dua S, Asu DE, Sarosha R, Kumar V (2006) Phytoremediation: cost effective approval for the removal of soil contaminants. In: Mukerji KG, Manoharachary C (eds) Current concepts in botany. I K International Publishing House, New Delhi, pp 425–446

Eggen T, Majcherczyk A (1998) Removal of polycyclic aromatic hydrocarbons (PAH) in con- taminated soil by white-rot fungus Pleurotus ostreatus. Int Biodeterior Biodegrad 41:111–117

Eggen T, Sveum P (1999) Decontamination of aged creosote polluted soil: the influence of tem- perature, white-rot fungus Pleurotus ostreatus, and pre-treatment. Int Biodeterior Biodegrad 43:125–133

Elekes CC, Busuioc G (2010) The Mycoremediation of metals polluted soils using wild growing species of mushrooms. Lat Trends Eng Educ:36–39

Eskander SB, Abd E-ASM, El-Sayaad H, Saleh HM (2012) Cementation of bioproducts gener- ated from biodegradation of radioactive cellulosic-based waste simulates by mushroom. ISRN Chem Eng. https://doi.org/10.5402/2012/329676

Gadd G (2001) Fungi in bioremediation. Cambridge University Press, Cambridge

Gaitán-Hernández R, Esqueda M, Gutiérrez A, Sánchez A, Beltrán-García M, Mata G (2006) Bioconversion of agrowastes by Lentinula edodes: the high potential of viticulture residues. Appl Microbiol Biotechnol 4:432–439

Hammel KE, Green B, Gai WZ (1991) Ring fission of anthracene by a eukaryote. Proceedings of the National Academy of Sciences 88(23):10605–10608

Haritash AK, Kaushik CP (2009) Biodegradation aspects of polycyclic aromatic hydrocarbons (PAHs): a review. J Hazard Mater 169:1–15

Heinfling A, Martinez MJ, Martinez AT, Bergbauer M, Szewyk U (1998) Transformation of industrial dyes by manganese peroxidases from Bjerkandera adusta and Pleurotus eryngii in a manganese-independent reaction. Applied and Environmental Microbiology 64:2788–2793

Isikhuemhen OS, Anoliefo G, Oghale O (2003) Bioremediation of crude oil polluted soil by the white-rot fungus, Pleurotus tuber-regium (Fr) Sing. Environ Sci Pollut Res 10:108–112

Jagtap VS, Sonawane VR, Pahuja DN, Rajan MG, Rajashekharrao B, Samuel AM (2003) An effec- tive and better strategy for reducing body burden of radiostrontium. J Radiol Prot 23:317–326

Jang KY, Cho SM, Seok SJ, Kong WS, Kim GH, Sung JM (2009) Screening of biodegradable function of indigenous ligno-degrading mushroom using dyes. Mycobiology 4:53–61

Jibran AK, MilseeMol JP (2011) *Pleurotus sajor-caju* Protein: a potential biosorptive agent. Advanced Biotech 4:25–27

Jonathan SG, Fasidi IO, Ajayi AO, Adegeye O (2008) Biodegradation of Nigerian wood wastes by P*leurotus tuber-regium* (Fries) Singer. Bioresour Technol 4:807–811

Johannes C, Majcherezyk A, Hutterman A (1996) Degradation of anthracene by lacasse of Trametes versicolor in the presence different mediator compounds. Applied Microbiology and Biotechnology 46:313–317

Kuforiji OO, Fasidi IO (2008) Enzyme activities of *Pleurotus tuber-regium* (Fries) Singer, cultivated on selected agricultural wastes. Bioresour Technol 4:4275–4278

Kuforiji OO, Fasidi IO (2009) Biodegradation of agro-industrial wastes by an edible mushroom *Pleurotus tuber-regium* (Fr.). J Environ Biol 4:355–358

Kulshreshtha S, Mathur N, Bhatnagar P, Jain BL (2010) Bioremediation of industrial wastes through mushroom cultivation. J Environ Biol 4:441–444

Kulshreshtha S, Mathur N, Bhatnagar P (2013) Fungi as Bioremediators: soil biology. In: Goltapeh EM, Danesh YR, Varma A (eds) Mycoremediation of paper, pulp and cardboard industrial wastes and pollutants. Springer, Berlin, Heidelberg, pp 77–116

Kulshrestha A, Mathur N, Bhatnagar P (2014) Mushroom as a product and their role in mycoremediation. AMB Express 4:29

Lamrood PY, Ralegankar SD (2013) Biosorption of Cu, Zn, Fe, Cd, Pb and Ni by non-treated biomass of some edible mushrooms. Asian J Exp Biol Sci 4:190–195

Lechner BE, Papinutti VL (2006) Production of lignocellulosic enzymes during growth and fruiting of the edible fungus *Lentinus tigrinus* on wheat straw. Process Biochem 4:594–598

Leonardi V, Vaclav Sasek V, Petruccioli M, D'Annibale A, Erbanova P, Cajthaml T (2007) Bioavailability modification and fungal biodegradation of PAHs in aged industrial soils. Int Biodeter Biodegr 60(3):165–170

Lin JE, Wang HY, Hickey RF (1990) Degradation kinetics of pentachlorophenol by Phanerochaete chrysosporium. Biotechnology and Bioengineering 35(11):1125–1134

Loske D, Huttermann A, Majerczk A, Zadrazil F, Lorsen H, Waldinger P (1990) Use of white rot fungi for the clean-up of contaminated sites. In: Coughlan MP, Collaco (eds) Advances in biological treatment of lignocellulosic materials. Elsevier, London, pp 311–321

Luo D, Yf X, Tan ZL, Li XD (2013) Removal of Cu^{2+} ions from aqueous solution by the abandoned mushroom compost of *Flammulina velutipes*. J Environ Biol 4:359–365

Matsubara M, Lynch JM, DeLeij FAAM (2004) A simple screening procedure for selecting fungi with potential for use in the bioremediation of contaminated land. www.aseanbiodiversity.info/Abstract/51006383.pdf. Accessed 24th July 2017

Morgan P, Lewis ST, Watkinson RJ (1991) Comparison of abilities of white-rot fungus to mineralise selective xenobiotic compounds. Appl Microbiol Biotechnol 34:693–696

Nagy B, Măicăneanu A, Indolean C, Mânzatu C, Silaghi-Dumitrescu MC (2013) Comparative study of Cd(II) biosorption on cultivated *Agaricus bisporus* and wild *Lactarius piperatus* based biocomposites. Linear and nonlinear equilibrium modelling and kinetics. *J Taiwan Inst Chem Eng*. https://doi.org/10.1016/j.jtice.2013.08.013

Nigam P, Banat IM, McMullan G, Dalel S, Marchant R (1995) Microbial degradation of textile effluent containing Azo, Diazo and reactive dyes by aerobic and anaerobic bacterial and fungal cultures, 37–38. Paper presented in 36th Annu. Conf. AMI, Hisar

Novotny C, Svobodova K, Erbanova P, Cajthaml T, Kasinath A, Lange E, Sasek V (2004) Ligninolytic fungi in bioremediation: extracellular enzyme production and degradation rate. Soil Biol Biochem 36(10):1545–1551

Okparanma RN, Ayotamuno JM, Davis DD, Allagoa M (2011) Mycoremediation of polycyclic aromatic hydrocarbons (PAH)-contaminated oil-based drill-cuttings. Afr J Biotechnol 10(26):5149–5156

Ollikka P, Alhonmaki K, Leppanen VM, Glumoff T, Raijola T, Suominen I (1993) Decolorization of azo, triphenylmethane, heterocyclic, and polymeric dyes by lignin peroxidase isoenzymes from Phanerochaete chrysosporium. Applied and Environmental Microbiology 59:4010–4016

Olusola SA, Anslem EE (2010) Bioremediation of a crude oil polluted soil with *Pleurotus Pulmonarius* and *Glomus Mosseae* using *Amaranthus Hybridus* as a test plant. J Bioremed Biodegr 4:111

Ouzouni PK, Petridis D, Koller W-D, Riganakos KA (2009) Nutritional value and metal content of wild edible mushrooms collected from West Macedonia and Epirus, Greece. Food Chem 115:1575–1580

Oyetayo VO, Adebayo AO, Ibileye A (2012) Assessment of the biosorption potential of heavy metals by *Pleurotus tuber-regium*. Int J Adv Biol Res 4:293–297

Pletsch M, De Araujo BS, Charlwood BV (1999) Novel biotechnological approaches in environmental remediation research. Biotechnol Adv 17(8):679–687

Pozdnyakova NN (2012) Involvement of the ligninolytic system of white-rot and litter-decomposing fungi in the degradation of polycyclic aromatic hydrocarbons. Biotechnol Res Int www.ncbi. nlm.nih.gov/pubmed/22830035. Accessed 24th July 2017

Purnomo AS, Mori T, Putra SR, Kondo R (2013) Biotransformation of heptachlor and heptachlor epoxide by white-rot fungus *Pleurotus ostreatus*. Int Biodeterior Biodegrad 4:40–44

Rajput Y, Shit S, Shukla A, Shukla K (2011) Biodegradation of malachite green by wild mushroom of Chhatisgrah. J Exp Sci 4:69–72

Rani P, Kalyani N, Prathiba K (2008) Evaluation of lignocellulosic wastes for production of edible mushrooms. Appl J Biochem Biotechnol 4:151–159

Sack U, Gunther T (1993) Metabolism of PAH by fungi and correction with extracellular enzymatic activities. J Basic Microbiol 33:269–277

Sasek V (2003) Why mycoremediations have not yet come into practice. In: The utilization of bioremediation to reduce soil contamination: problems and solution. Kluwer Academic Publishers, Amsterdam, pp 247–266

Sasek V, Cajthaml T (2005) Mycoremediation. Current state and perspectives. Int J Med Mushrooms 7(3):360–361

Sesli E, Tuzen M (1999) Level of trace elements in the fruiting bodies of macrofungi growing in the east black sea region of Turkey. Food Chem 65:453–460

Singh H (2006) Mycoremediation: fungal bioremediation. Wiley-Interscience, New York

Stamets PE (2010) *Mycoremediation and Its Applications to Oil Spills*. www.realitysandwich. com/mycoremediation_and_oil_spill

Stella T, Covino S, Křesinová Z, D'Annibale A, Petruccioli M, Cajthaml T (2012) Mycoremidiation of PCBs dead-end metabolites: *In vivo* and *In vitro* degradation of chlorobenzoic acids by the white rot fungus *Lentinus tigrinus*. Environ Eng Manag J 11(3)

Sutherland C, Venkobachar C (2013) Equilibrium modeling of Cu (II) biosorption onto untreated and treated forest macro-fungus *Fomes fasciatus*. Int J Plant Anim Environ Sci 4:193–203

Sykes C (2002) Magical Mushrooms: Mycoremediation. www.realitysandwich.com/ mycoremediation_and_oil_spills

Tanaka H, Itakura S, Enoki A (1999) Hydroxyl radical generation by an extracellular low- molecular–weight substance and phenol oxidase activities during wood degradation by the white–rot basidiomycetes *Trametes versicolor*. J Biotechnol 75(1):57–70

Tay CC, Liew HH, Yin CY, Abdul-Talib S, Surif S, Suhaimi AA, Yong SK (2011) Biosorption of cadmium ions using *Pleurotus ostreatus*: growth kinetics, isotherm study and biosorption mechanism. Korean J Chem Eng 4:825–830

Thakur M (2014) Mycoremediation—a potential tool to control soil pollution. Asian J Environ Sci 9(1):24–31

Thakur M (2015) Wild mushrooms as natural untapped treasures. In: Chauhan AK, Pushpangadan P, George V (eds) Natural products: recent advances. Write & Print Publications, New Delhi, pp 214–226

Thomas SA, Aston LM, Woodruff DL, Cullinan VI (2009) Field demonstration of Mycoremediation for removal of Fecal coliform Bacteria and nutrients in the Dungeness watershed, Washington. Pacific Northwest National Laboratory, Richland, Washington

Tismal M, Zelic B, Vasic-Racki D (2010) White-rot fungi in phenols, dyes and other xenobiotics treatment – a brief review. Croatian J Food Sci Technol 2(2):34–47

Tsujiyama S, Muraoka T, Takada N (2013) Biodegradation of 2,4-dichlorophenol by shiitake mushroom (*Lentinula edodes*) using vanillin as an activator. Biotechnol Lett 4:1079–1083

Vanaken B, Godefroid L, Peres C, Naveau H, Agathos S (1999) Mineralization of C-U-ring labeled 4-hydroxylamino-2,6-dinitrotoluene by manganese-dependent peroxidase of the white-rot basidiomycete. Journal of Biotechnology 68(2-3):159–169

Watanabe K (2001) Microorganisms relevant to bioremediation. Curr Opin Biotechnol 12:237–241

Williams RT, Ziegenfuss PS, Sisk WE (1992) Composting of explosives and propellant contaminated soils under thermophilic and mesophilic conditions. J Ind Microbiol Biotechnol 9(2):137–144

Zhu MJ, Du F, Zhang GQ, Wang HX, Ng TB (2013) Purification a laccase exhibiting dye decolorizing ability from an edible mushroom *Russula virescens*. Int Biodeterior Biodegrad 4:33–39

Establishing Correlation Between Abiotic Stress and Isoprene Emission of Selected Plant Species

Pallavi Saxena and Chirashree Ghosh

Abstract The atmospheric hydrocarbon budget influenced by a vast range of volatile organic compounds (VOCs) is both anthropogenic and biogenic in origin. Evolution of plant VOCs is a complex process affected by interactions of plants with biotic and abiotic factors in constantly changing environment but their functional role is still a matter of speculation. Isoprene (2-methyl 1,3-butadiene), a five-carbon hydrocarbon, is emitted from the leaves of many plant species. In the present study, isoprene emission potential of two urban plant species (*Dalbergia sissoo* and *Nerium oleander)* were measured using branch enclosure method at different selected sites on the basis of proximity to traffic density. In order to find out the dependence of isoprene emission with abiotic factors (temperature and photosynthetic active radiation PAR), regression analysis has been performed. In the case of *Dalbergia* sp. high temperature and PAR promote high isoprene emission during summer months. Thus, positive linear relationship gives the best fit between temperature, PAR, and isoprene emission rate during summer season as compared to other seasons, whereas in the case of *Nerium sp.*, no such appropriate relationship was obtained. The study concludes that *Dalbergia sissoo* comes under high isoprene emission category, while *Nerium oleander* comes under BDL (below detection limit) variety. For any greenbelt development program, it is very important to select the type of plant species to be planted. The present small study reflects that *Nerium oleander* should be planted in the outskirts of selected areas and that planting of *Dalbergia sissoo* should be done on low scale so that the air remains clean and indirect production of tropospheric ozone, aerosol production will be minimized.

Keywords Isoprene · Temperature · PAR · Roadside · VOCs and air quality

P. Saxena (✉)
Department of Environmental Sciences, Hindu College, University of Delhi, Delhi, India

C. Ghosh
Environmental Pollution Laboratory, Department of Environmental Studies,
University of Delhi, Delhi, India

© The Author(s), under exclusive licence to Springer Nature Switzerland AG 2019   43
T. Jindal, *Emerging Issues in Ecology and Environmental Science*,
SpringerBriefs in Environmental Science, https://doi.org/10.1007/978-3-319-99398-0_4

1 Introduction

Volatile organic compounds (VOCs) are defined as organic compounds having a vapour pressure greater than 10^{-1} Torr at 25 °C and 760 mmHg (USEPA 1999). Many such compounds are released by plants are normally difficult for humans to detect (Font et al. 2011). Biogenic VOCs are involved in a range of ecological functions, including indirect plant defense against insects (Bezemer and van Dam 2005; Mithofer and Boland 2012), pollinator attraction (Hewitt 2009), plant–plant communication (Peñuelas and Llusia 2003; Iriti and Faoro 2009), plant–pathogen interactions (Tao and Jain 2005), reactive oxygen species removal (Loreto and Schnitzler 2010), thermotolerance (Sharkey et al. 2008), and other environmental stress adaptations (Dudareva et al. 2005). Their evolution is complex, affected by interactions of plants with biotic and abiotic factors in constantly changing environments (Pichersky et al. 2006). Whether plant VOCs evolved mainly as a defence against biotic and abiotic stress or to serve plant reproduction as a means of attracting pollinators and seed dispersers is still a matter of speculation (Chen et al. 2011). However, VOC production and emission can be affected by abiotic factors, such as temperature and light, which changes the profile of carbon dioxide (CO_2) and ozone (O_3). The biogenic non-methane VOC emission pattern (based on their stability) affects atmospheric reactions, contributing to sometimes secondary organic aerosol formation during plant-pest defence interaction (Tao and Jain 2005).

Isoprene (2-methyl 1, 3-butadiene), a five-carbon hydrocarbon, is emitted from the leaves of many plant species. Its annual global emission is estimated at 500 Tg C year^{-1} from vegetation to the atmosphere (Guenther et al. 1995), which is equivalent to total methane emission (Brilli et al. 2007). In addition to isoprene, many plant species emit significant amount of monoterpenes. In the atmosphere, VOC rapidly reacts with OH· radical and indirectly with NO_x through a complex series of reactions to produce a variety of compounds, including ozone, carbon monoxide, carbonyls, organic nitrates, peroxyacetylnitrate, organic acids, and atmospheric aerosols (Possell et al. 2005). Depending upon ambient nitrogen oxides (NO_x) concentration, isoprene oxidation leads to either production or consumption of ground level ozone (Chameides et al. 1988).

It has long been known that isoprene emission is highly temperature and light dependent (Sanadze and Kalandaze 1966; Tingey et al. 1979; Monson et al. 1992). Isoprene emission increases up to 35 °C to 40 °C even when carbon assimilation is declining. This uncoupling of emission from photosynthesis contributed to the hypothesis that isoprene may protect plants against heat stress (Sharkey and Singsaas 1995; Singsaas et al. 1997). The rate of isoprene emission declines above its optimum, but the optimum temperature is significantly affected by the protocol of emission measurement (Singsaas et al. 1999; Singsaas and Sharkey 2000). If measurements are made quickly, the optimum is much higher than the measurements which are made slowly. This occurs because isoprene emission above 35 °C is unstable, increasing when the temperature first rises but then falls back after

10–20 min at the higher temperature. A mechanistic understanding of the regulation of isoprene emission with changes in temperature is very important to model accurately isoprene output in future environments where global mean temperature is predicted to rise. In the case of light, short-term (up to 20 min) effects of light intensity on isoprene emission rates reflect that the leaves which develop in full sun emit isoprene at a higher rate than the leaves that develop in shade (Sharkey et al. 1991; Harley et al. 1994).

Massive plantation or afforestation programs have been initiated since 1979 to increase vegetation cover in India (SFR 2005). Under these programs, a variety of plant species (trees, shrubs and herbs) have been planted in both urban and rural areas of India. Greenbelt development is one of the popular air quality mitigation programs which influence lowering of air temperature, soil fertility improvement and sequestration of CO_2. It is reported by various scientists that many plant species have been found to emit highly reactive isoprene (Rasmussen 1972; Tingey et al. 1979; Karlik and Winer 2001; Harley et al. 2004; Köksal et al. 2010). Isoprene emission is species specific, varying as much as four orders of magnitude depending upon the plant species (Benjamin et al. 1996). So, large scale planting of high emitting plant species is associated with potential air quality liability, particularly in polluted urban air sheds. In view of this, it is important to select low emitting plant species for plantation programs. Till now, in our country, isoprene emission potential of plant species is not taken in to consideration while selecting plant species for greenbelt development program, probably due to availability of limited information. In the present study, isoprene emission capacity at the bottom of the canopies of *Dalbergia sissoo* and Nerium oleander at different sites selected on the basis of land use pattern, viz., near to traffic intersection with dense vegetation (Site RZ1: University of Delhi, DU), away from traffic intersection with dense vegetation under floodplain area (Site RZ2: Yamuna Biodiversity Park, YBP) and away from traffic intersection with dense vegetation under hilly ridge area (Site RZ3: Aravalli Biodiversity Park,ABP) during three different seasons (monsoon, winter, and summer) in Delhi were measured. In order to find out the dependence of isoprene emission rate on enclosure temperature and photosynthetic active radiation (PAR), regression analysis has been performed and linear and second degree fitting had been tried out depending on the best fit conditions.

2   Materials and Methods

2.1   Sites Description

The capital city of India, Delhi, situated on the banks of Yamuna River, is one of its largest cities which lies at an altitude of between 700 and 1000 ft, with an area of approximately 1500 km². Delhi has a tropical steppe climate with continental air leading to relatively dry conditions and extremely hot summers. Monthly mean temperature ranges from 14.3 °C in January (minimum 3 °C) to 34.5 °C in June (maximum 47 °C) and the annual mean temperature is 25.3 °C. There are three main seasons in Delhi, viz., monsoon, winter, and summer. The mean annual total rainfall is 715 mm. Heavy rains of the monsoon act as a cleansing agent which wash out high concentration of atmospheric pollutants. Wind speeds are typically higher during summer and monsoon months than in winter. Based on a recent report by Goyal and Khaliq 2011, Delhi is among the ten most polluted cities in the world. Population and transport are the main reasons behind the rising concentrations of air pollutants in Delhi. Creation of green belts has been identified as one of the most cost-effective air pollution abatement methods. The flora in Delhi largely consists of some common trees like Azadirachta indica, Ficus religiosa, *Mangifera indica, and Eucalyptus*. Besides these, other common ornamental plants and shrubs (*Dracaena deremensis* (Family:Dracaenaceae), *Lantana indica*, Lantana camara, *Bambusa indica*, Tagetes erecta (Family: Asteraceae), Rosa indica (Family: Rosaceae), Dianthus caryophyllus (Family:Caryophyllaceae), Petunia hybrid*a, etc.*) are planted in small adjacent gardens near different emitting zones, such as a residential, commercial, industrial, or traffic intersection (Saxena and Ghosh 2013).

Sampling sites were selected on the basis of land use pattern, viz., near to traffic intersection with dense vegetation (Site RZ1: University of Delhi, DU; 500 m from traffic zone), away from traffic intersection with dense vegetation under floodplain area (Site RZ2: Yamuna Biodiversity Park, YBP; 1.5 km away from traffic zone) and away from traffic intersection with dense vegetation under hilly ridge area (Site RZ3: Aravalli Biodiversity Park,ABP; 2 km away from traffic zone) during three different seasons (monsoon, winter, and summer) in Delhi in the year 2011.

2.2   Plant Material

Two commonly occurring plant species, i.e., *Dalbergia sisso* and Nerium oleander were selected for the study on the basis of their wide abundance, local availability, and representation of certain families and genera. *D. sissoo* and N. oleander have the property to shed their leaves in Jan–Feb (winter season) and then gain new leaves from March onwards. Summer season (Apr–June) is considered to be the best for their abundant growth.

2.3   Selection of Seasons

These commonly occurring plant species were sampled at three selected sites during three different seasons in 2011, viz., summer (Apr–June), monsoon (Aug–Sept), and winter (Nov–Dec). Sampling was done for 8 h from 10:00 to 17:00 h consecutively for 3 days for each plant species at each site in every selected month.

During the winter season (Nov–Feb), sampling was done during November and December only since these plants shed leaves and are left with hardly any leaves after these months of winter.

2.4 Isoprene Measurement

The composition of volatile emissions is usually quantified by the analysis of air samples collected in glass or plastic containers in which branches of living plants are placed (Zimmerman et al. 1978; Knoppel et al. 1981). This method is known as branch enclosure method (Zimmerman et al. 1978). The end of a branch of a tree was carefully introduced into a glass cylinder of 800 mm (diameter). Then it was closed with a foam plastic plug cut into two parts and sample collection was started immediately. The outer end was connected to a sorption tube (250×6 mm) packed with 0.6–0.7 g of Chromosorb. The air passing through an inlet in the plug was drawn through the tube and the sorbent layer at a rate of 0.5 l min^{-1} with the help of an organic vapor sampler (OVS). The total sample volume was 4.78 l. The temperature was measured with a thermometer and light was measured with quantum sensor (Model No. SI: 121) located inside the cylinder. After the sampling had been finished, the leaves were separated and weighed. Desorption of the adsorbed isoprene was done using carbon disulfide (CS_2). CS_2 has the property to dissolve the gases properly which are adsorbed in chromosorb. The substances desorbed in the CS_2 were analyzed by capillary gas chromatography (Shimazdu, GC-2010), equipped with Supelcowax column. A flame ionization detector (FID) was used for analysis, while quantification was done using standards from Sigma-Aldrich. The mass spectra were recorded at 70 eV with accelerating voltage of 3.5 kV cathode current of 25 PA. The initial identification was carried out according to mass spectra and further identification was performed from the retention parameters of chromatographic peaks.

2.5 Measurement of Environmental Parameters

Temperature and Photosynthetic Active Radiation (PAR) were measured both outside and inside of experimental setup.

- **PAR**: PAR was measured by Apogee Quantum Meter (Model no. MQ-200) μmol/m^2/s after every 1 h. The sensor of the Quantum Meter was inserted inside the glass chamber and suitably oriented for measurement.
- **Temperature**: Temperature readings were taken after every 1 h in degree Celsius. The temperature inside the enclosure chamber was found to be relatively higher (approx. 2 °C).
- **Foliar mass determination**: After the emission flux measurements were complete, the entire branch enclosed in the chamber was harvested and the leaves were dried in an oven at 70 °C to a constant weight.

3   Results and Discussion

The isoprene emission rates were measured for two plant species, viz., *Dalbergia sissoo* and Nerium oleander at three different selected sites Sites RZ1 (DU), RZ2 (YBP), and RZ3 (ABP). The mean isoprene emission rates varied from 51.60 ± 0.43 μg/g/h – 60.14 ± 1.80 μg/g/h at selected sites in the case of D. sissoo while in the case of N. oleander, it varied from 0.02 ± 0.01 μg/g/h – 0.03 ± 0.01 μg/g/h. Maximum isoprene emission rate was observed in the case of D. sissoo as compared to N. oleander irrespective of sites. Moreover, significant variation of foliar mass was observed during different seasons in both the plant species. Significant decrease in foliar mass was noticed during winter season as compared to summer and monsoon, while nonsignificant variation was observed between summer and monsoon seasons. This is because during winter season, leaves were shed off in both the plant species, resulting in lower foliar mass estimation.

Plant species screened for isoprene emission in the present study may be grouped into four categories proposed by Karlik and Winer 2001, namely (1) Below Detectable Limit (BDL) isoprene emitting ($\leq$ 1 μg/g/h), (2) low emitting (1 $\leq$ to <10 μg/g/h), (3) moderate emitting (10 $\leq$ to <25 μg/g/h), and (4) high emitting ($\geq$ 25 μg/g/h). In the present study, Nerium oleander falls under BDL isoprene emitting category and Dalbergia sissoo, under high isoprene emitting category. Some progress has been made in explaining that "why some plants emit high or low isoprene". It has been assumed that the capacity for enzyme-catalyzed isoprene emission has evolved independently within distinct lineages of plants, and may have been lost from some lineages (Loreto et al. 1998; Harley et al. 1999; Sharkey et al. 2005). For example, Family Fabaceae has groups with high taxonomic diversity with numerous isoprene-emitting genera and the trait is distributed among traditionally defined subfamilies (Monson et al. 2012), and this statement is in accordance with our observations described above in the case of D. sissoo which comes under Family Fabaceae, whereas Family Apocynaceae has less taxonomic diversity, and the trait is not normally distributed like in our findings in the case of N. oleander. In addition to that, isoprene is synthesized by the action of IspS (isoprene synthase) on dimethylallyl diphosphate (DMADP) (Silver and Fall 1991) produced by the MEP pathway (Schwender et al. 1997). In plants which emit low emissions, it is more likely that non-functionalization of the this light-dependent IspS occur which ultimately is responsible for an inability to generate adequate DMADP substrate causing mutations in the methylerythritol 4-phosphate (MEP) pathway which tend to interfere with metabolic processes that are crucial to plant survival and ultimately emit less isoprene (Estévez et al. 2001; Fares et al. 2006; Rodriguez-Concepción 2010). According to Sharkey et al. (2008), tolerance of heat flecks can help explain the distribution of the capacity to emit isoprene among plants. The plants which have high stomatal conductance allow for high rates of latent heat loss, buffering against heat flecks and thus emit very less or no isoprene from them. Similar is the case of N. oleander in the present study, as these plants are subjected to intense evaporation of water, a resource which is often in short supply. Many such plants have a number of modifications which minimize water loss through transpiration, the evaporation of water from the plant surfaces. Some plants drop their leaves during periods of drought; cactus plants photosynthesize with modified stem tissues, and lack leaves entirely. Those plants which do produce and retain leaves often have special features which are associated with the xeromorphic leaf and this leads to high stomatal conductance. The property of high stomatal conductance indicates that stomata of *N. oleander* arranged in crypts with trichomes may be partially open (Lakušić et al. 2007; Roth-Nebelsick et al. 2009). Although the adaptive significance of stomatal encryption is still under debate, it has been shown that crypts facilitate CO_2 diffusion in the photosynthetic mesophyll of thick leaves (Hassiotou et al. 2009), while the presence of stomata on one surface increases the distance of CO_2 diffusion to the photosynthetic mesophyll cells. Thick mesophyll of this species (approximately 290 µm) enhances the overall gas exchange parameters as compared to leaves that possess a thinner mesophyll (approximately 178 µm) as in the case of the present study plant, *Dalbergia sp.* It has less stomatal conductance as compared to N. oleander which favors high isoprene emission rate. Moreover, D. sissoo is a deciduous tree and trees are generally the biggest isoprene emitters. In the tropics, plant leaves can grow very large, and this creates a large boundary layer insulating the leaf from air temperature, allowing the leaf temperature to exceed air temperature by 10 °C and more. Also, in humid air, heat loss by latent heat of evaporation is reduced. The humid tropics are known to have many isoprene-emitting species (Sharkey and Yeh 2001). Thus, there is a correspondence between the distribution of isoprene emission capacity among plant species and its presumed function in increasing tolerance of heat flecks suffered by leaves.

3.1 Effect of Temperature and Solar Radiation on Isoprene Emission in Dalbergia sissoo

According to Sharkey et al. (1996), at 30 °C approx., 2% and at 40 °C approx., 15% of carbon fixed during photosynthesis is emitted as isoprene. Enclosure chamber temperature was used in place of leaf temperature in absence of a reliable instrument for measuring this type of temperature. Previous studies have shown that branch enclosure temperature can be used in place of leaf temperature (Street et al. 1996; Owen and Peñuelas 2005). Generally, leaf temperatures are 2 °C lower than branch enclosure temperature (Owen et al. 1997).

   In the case of Dalbergia sissoo, isoprene emission was measured in three different seasons, viz., summer, winter, and monsoon at recorded enclosure temperatures and PAR at different selected sites. In order to find out the dependence of isoprene emission rate on enclosure temperature and PAR, regression analysis had been performed and linear and second degree fitting had been tried out depending on the best fit conditions. The scatter plot between isoprene emission rate and enclosure temperature and PAR along with fitting had been shown in Fig. 1a–c and Fig. 2a–c. It

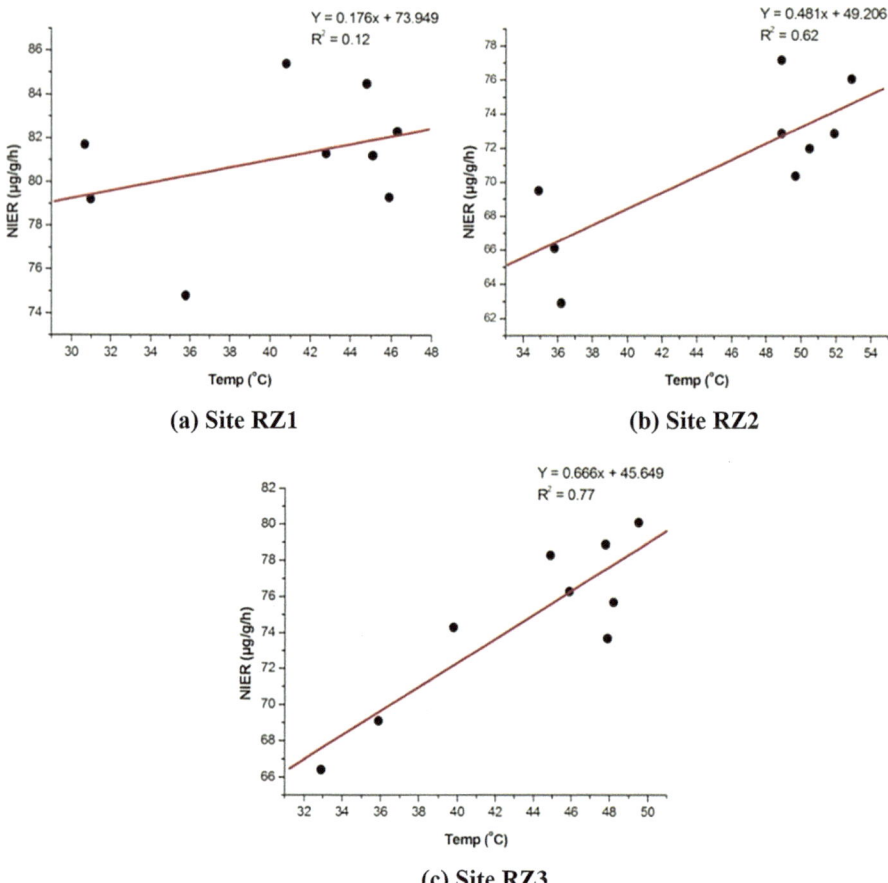

Fig. 1 (**a–c**) Variation of isoprene with enclosure temperature at different sites during summer season in *D. sissoo*

can be seen from the figures that for summer season, with increase in temperature and solar radiation (PAR), there is increase in isoprene emission rate which is in accordance with Guenther et al. (2006). Thus, positive linear relationship gives the best fit between temperature and isoprene emission rate [R^2 = 0.12, 0.62, and 0.77 at RZ1 (University of Delhi, DU), RZ2 (Yamuna Biodiversity Park, YBP), and RZ3 (Aravalli Biodiversity Park, ABP) respectively] and PAR and isoprene emission rate [R^2 = 0.29, 0.70, and 0.42 at RZ1, RZ2, and RZ3 respectively] during summer season as compared to winter and monsoon season. The high correlation coefficient between temperature and isoprene emission rate at sites RZ2 and RZ3 during summer season with weak correlation at site RZ1 suggests that there is some factor other than temperature which is responsible for the isoprene emission rate. On the other hand, at Site RZ2, high correlation was found in the case of PAR and isoprene emission rate while RZ1 and RZ3 showed weak correlation. In short, it is observed

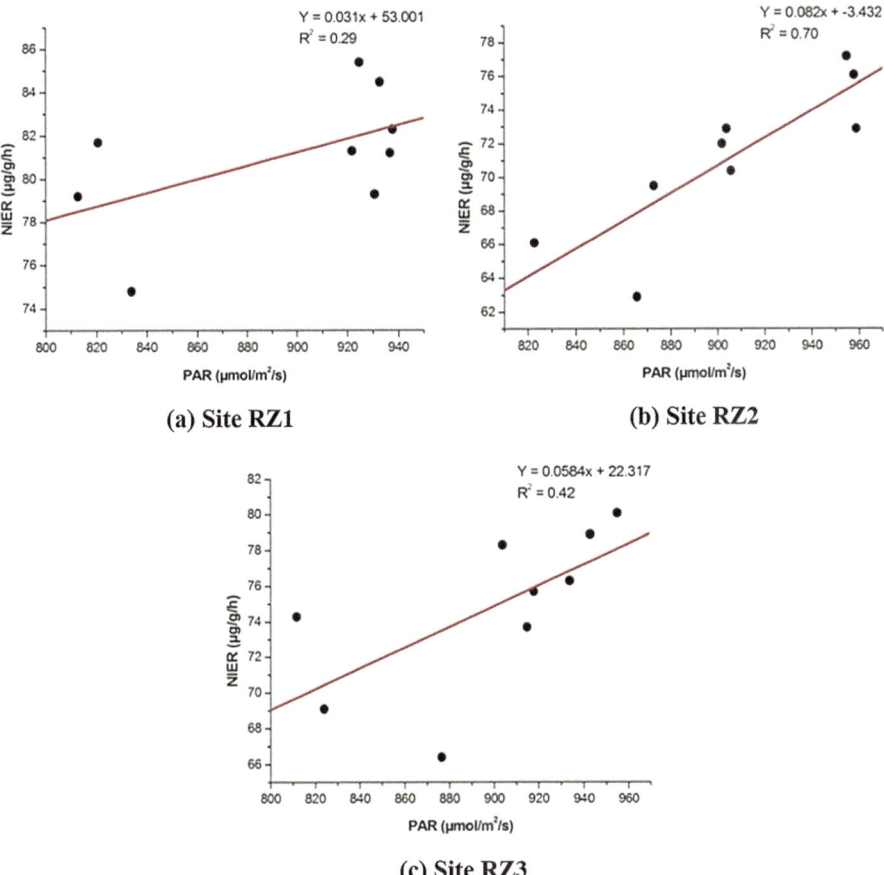

Fig. 2 (**a–c**) Variation of isoprene with PAR at different sites during summer season in *D. sissoo*

that during summer season, at Site RZ2 both temperature and PAR are highly responsible for isoprene emission rate. At RZ3, the role of temperature is higher than PAR while at RZ1 there are some other factors than temperature and PAR which are responsible for isoprene emission rate.

During winter season, at Site RZ1 and RZ2, isoprene emission rate decreases with increase in temperature while at RZ3 isoprene emission rate first increases and then decreases with increasing temperature. Thus, during winter season, second degree polynomial fitting has been found out between temperature and isoprene emission rate with R^2 of about 0.77 and 0.89 for Site RZ1 and RZ3, respectively. On the other hand, linear fitting with weak correlation ($R^2 = 0.25$) has been found out at Site RZ2 (Fig. 3a–c). In the case of PAR, second degree polynomial fitting gives the best fit for site RZ1 with high R^2 of about 0.89. On the other hand, at sites RZ1 and RZ2, linear fitting gives the best fit with the R^2 of about 0.64 and 0.33 (Fig. 4a–c). Similarly, during monsoon season, second degree polynomial fits best between tem-

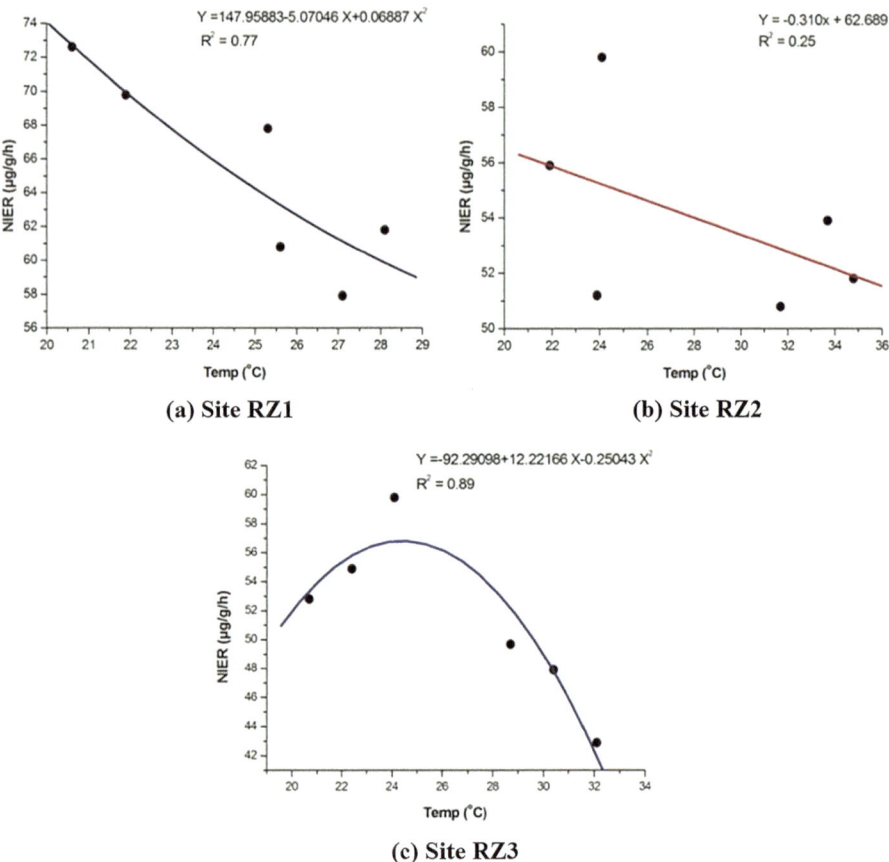

Fig. 3 (**a–c**) Variation of isoprene with enclosure temperature at different sites during winter season in *D. sissoo*

perature with isoprene emission rate at site RZ1 and RZ3 with weaker correlation (R^2 = 0.20 and 0.30 at site RZ1 and RZ3 respectively) while linear fitting gives the best fits with very weak correlation (R^2 = 0.02) at Site RZ2 (Fig. 5a–c). In the case of PAR, second degree polynomial fits best with moderate correlation having R^2 of about 0.55 and 0.51 at Site RZ1 and RZ3, respectively, and high R^2 of about 0.80 at Site RZ2. At all sites, isoprene emission rate shows negative linear relationship with temperature as well as PAR (Fig. 6a–c). Thus, during monsoon season, second degree polynomial fitting gives the best fitting for all sites except at site RZ2 (Yamuna Biodiversity Park, YBP) where linear fitting gives the best fit, both in the case of temperature and PAR (Figs. 5 and 6). For winter season, second degree fitting gives the best fit for site RZ1 (University of Delhi, DU) and RZ3 (Aravalli Biodiversity Park,ABP) in the case of enclosure temperature and RZ1 in the case of PAR. For all the rest cases of winter season, linear fitting gives the best fit as can be seen from Figs. 3 and 4. During winter and monsoon season, both temperature and

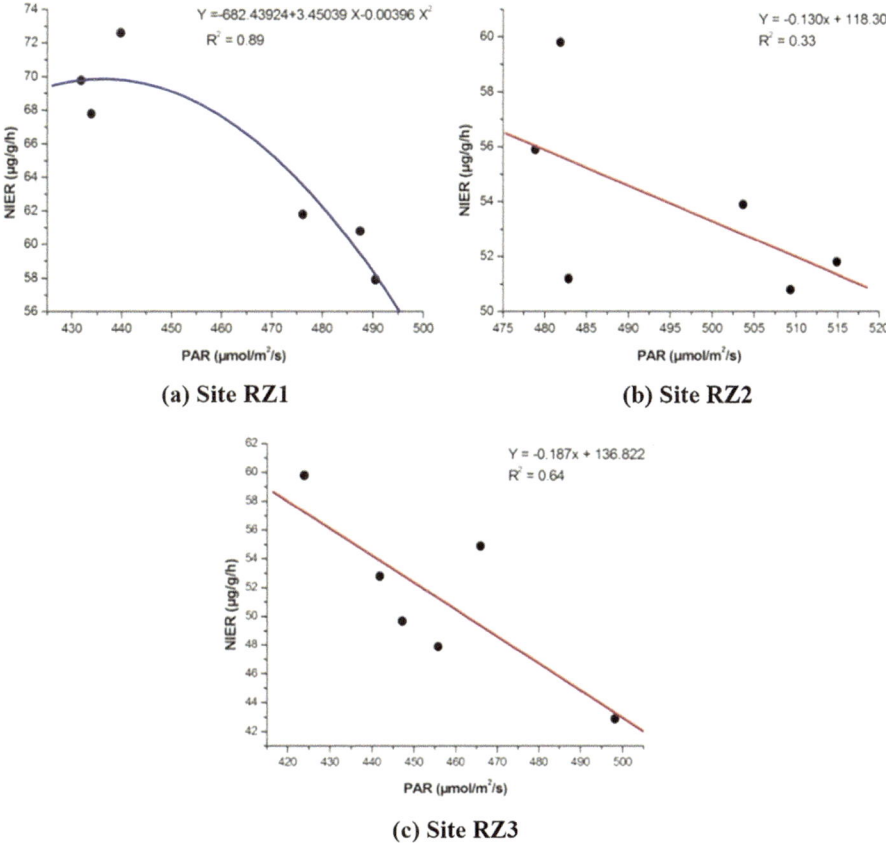

Fig. 4 (**a–c**) Variation of isoprene with PAR at different sites during winter season in *D. sissoo*

PAR show poor and negative linear result of the regression fitting with isoprene emission rate which is not in accordance with Guenther et al. (2006). This could be because of the fact that during winter season, leaves were mostly shed off from D. sissoo which in turn leads to less isoprene emission. On the other hand, during monsoon, due to low temperature and light as compared to summer season less isoprene emission rate was found.

The above observations depict that due to high temperature and PAR, high isoprene emission rates were found during summer months as compared to winter and monsoon seasons where decreased temperature and PAR were reported. This might be due to the reason that, as per thermotolerance hypothesis (Sharkey and Singsaas 1995; Sharkey 1996), plants emit isoprene to protect the photosynthetic apparatus from damage caused by high temperature. According to Monson et al. (1992) and Schnitzler et al. (1996), isoprene emission increases with increasing temperature because isoprene synthesis is related to activity of isoprene synthase enzyme which is highly temperature sensitive.

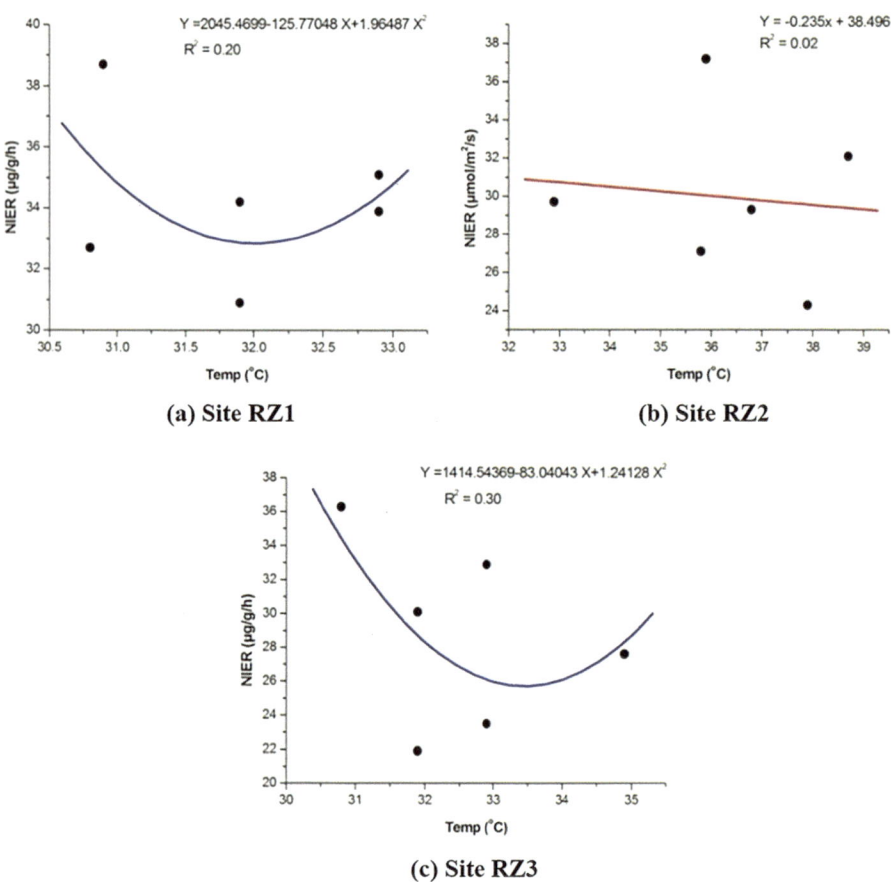

Fig. 5 (a–c) Variation of isoprene with PAR at different sites during monsoon season in *D. sissoo*

3.2 *Effect of Temperature and Solar Radiation on Isoprene Emission in Nerium oleander*

As already discussed, N. oleander emits negligible amount of isoprene which comes under below detectable limit category. Therefore, no such appropriate relationship was obtained between isoprene emission rate and temperature and PAR for all the considered seasons and sites as can be easily identified from the Figs. 1, 2, 3, 4, 5 and 6. During summer season, for sites RZ1 (University of Delhi, DU) and RZ3 (Aravalli Biodiversity Park, ABP), the isoprene emission rate first increases with increasing temperature and then begins to decrease at higher

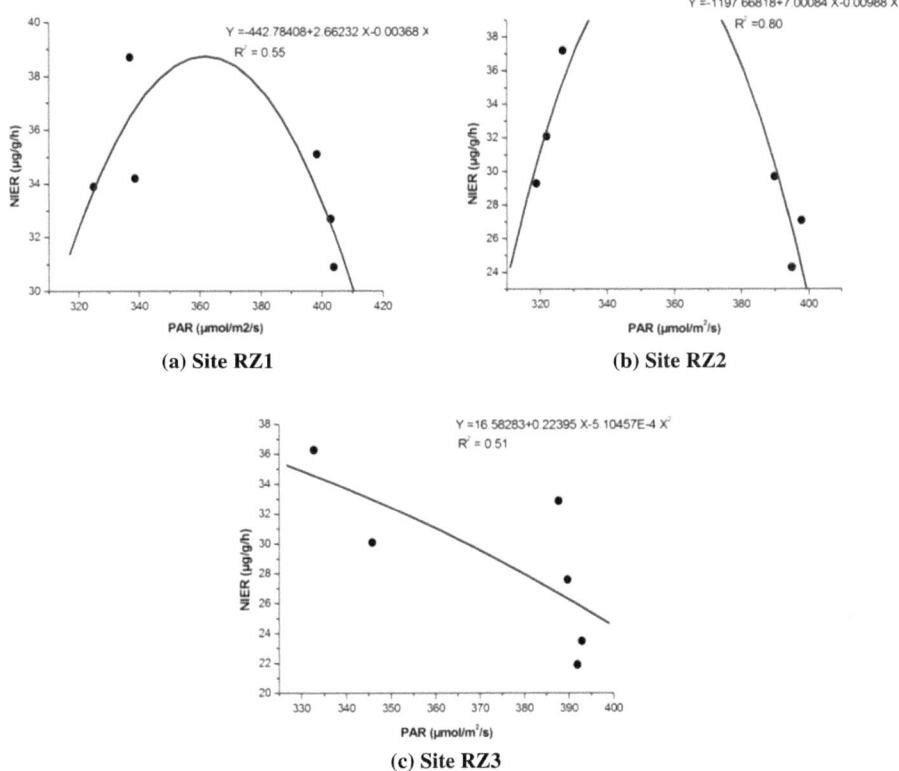

Fig. 6 (**a–c**) Variation of isoprene with PAR at different sites during monsoon season in *D. sissoo*

temperatures, whereas at site RZ2 (Yamuna Biodiversity Park, YBP), the isoprene emission rate, first decreases and then increases with increasing temperature. Thus, the second degree polynomial fitting gives the best fit between the isoprene emission rate and temperature with the R^2 of about 0.57, 0.13, and 0.19 for sites RZ1, RZ2, and RZ3, respectively (Fig. 7a–c). In relation with PAR during the summer season, second degree polynomial fitting gives the best fit for RZ1 and RZ3 with the weak correlation, i.e., R^2 of about 0.17 and 0.23 for RZ1 and RZ3, respectively (Fig. 8a–c). During winter season, the isoprene emission rate appears to be increasing with increasing temperature for site RZ1 and RZ2. Thus, positive linear fitting gives the best fits with the moderate correlation ($R^2 = 0.65$) and weak correlation ($R^2 = 0.17$) for site RZ1, RZ2, and RZ3, respectively. The second degree polynomial fitting has been obtained for site RZ3 with good correlation ($R^2 = 0.88$) (Fig. 9a–c). In relation with PAR, linear fits has been obtained for site RZ1 with the

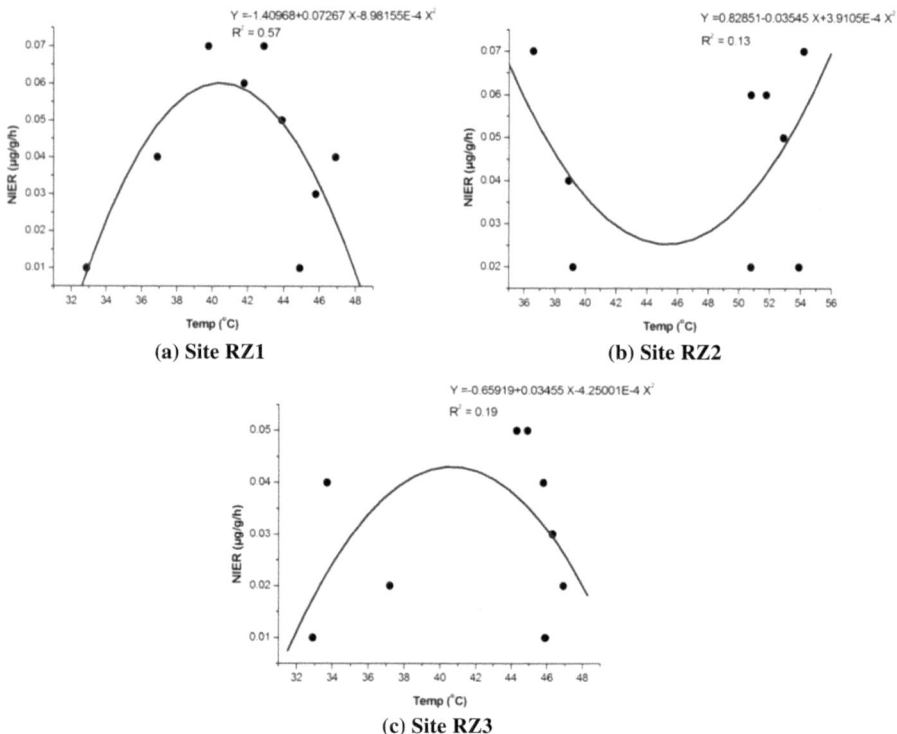

Fig. 7 (**a–c**) Variation of isoprene with enclosure temperature at different sites during summer season in *N. oleander*

moderate correlation ($R^2 = 0.53$) and second degree polynomial fitting gives the best fits for site RZ2 and RZ3 with weak correlation (i.e., R^2 of about 0.33 and 0.25 for RZ2 and RZ3 respectively) (Fig. 10a–c). For monsoon season, the isoprene emission rate shows different dependence for different sites as can be seen from the Figs. 11 and 12. At site, RZ1, the isoprene emission rate first decreases and then increases with increasing temperature and thus, second degree polynomial fitting gives the best fit with moderate correlation ($R^2 = 0.63$). At site, RZ2, the isoprene emission rate decrease with increasing temperature and thus, negative linear relationship has been obtained with moderate correlation ($R^2 = -0.67$). At site, RZ3, the isoprene emission rate increases with increasing temperature and thus, linear relationship has been obtained with moderate correlation ($R^2 = 0.65$) (Fig. 11a–c). On the other hand, the isoprene emission rate shows very weak correlation with PAR for monsoon season. The linear relationship with very weak correlation has been obtained for site RZ1 and RZ3 having R^2 of about 0.03 and 0.04, respectively. On the other hand, second degree relationship has been obtained for site RZ2 with the R^2 of about 0.33 (Fig. 12 (a–c)).

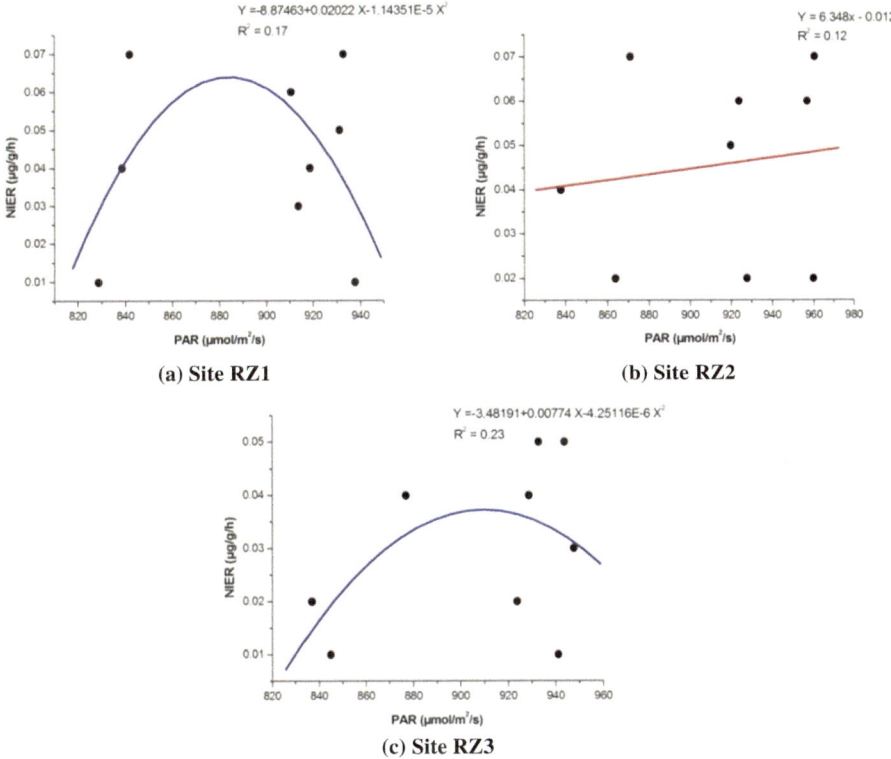

Fig. 8 (**a–c**) Variation of isoprene with PAR at different sites during summer season in *N. oleander*

In short, it has been clearly noted that seasonally there is no linear positive significant dependence of temperature and solar radiation (PAR) with isoprene emission rate in all the seasons in the case of N. oleander. As a result N. oleander emits negligible amount of isoprene without any dependence on temperature and solar radiation (PAR).

3.3 Calculation of Isoprene Emission Rate of Selected Plant Species at Particular Sites

$$E_{ISO} = n \times Es_{(D.sissoo)} C_L C_T km^2_{(Site)} + n \times Es_{(N.oleander)} C_L C_T km^2_{(Site)}$$

Where, E_{ISO} = Total isoprene emission rate, n = number of plants in each species, Es = isoprene emission rate of particular plant species, C_L & C_T = respectively light and temperature coefficient, km^2 (Site) = area of each site (Tables 1, 2, 3 and 4).

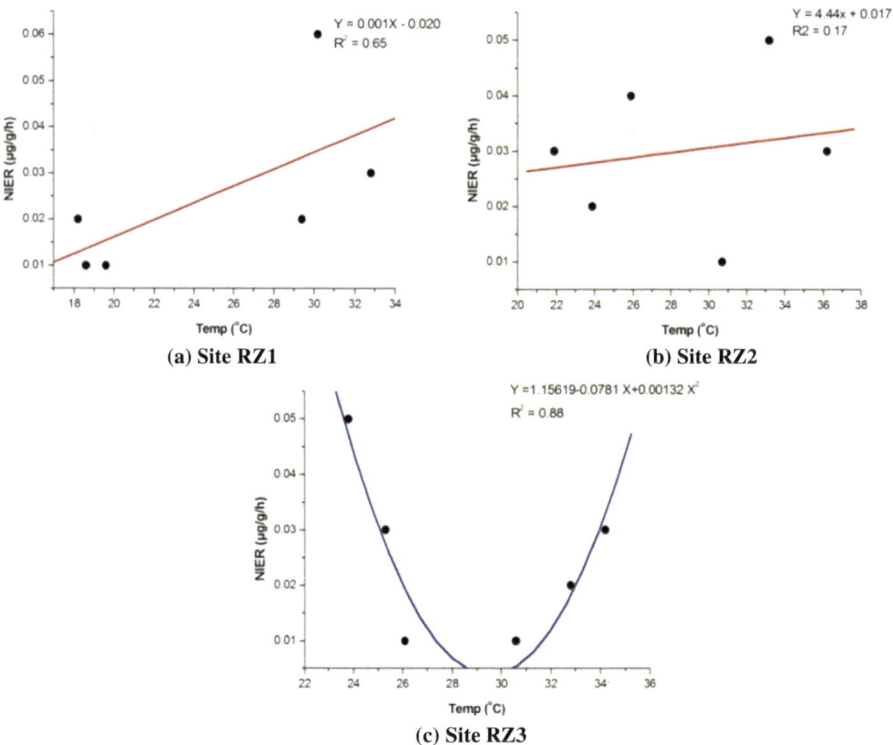

Fig. 9 (**a–c**) Variation of isoprene with enclosure temperature at different sites during winter season in *N. oleander*

The estimation of total isoprene emission rate per area/per site can be calculated as per the above formula given by Westberg et al. (2000).

- E_{ISO} of Site DU (RZ1) = 434.87 µg/g/h.
- E_{ISO} of Site YBP (RZ2) = 1213.76 µg/g/h.
- E_{ISO} of Site ABP (RZ3) = 706.54 µg/g/h.

From the above data, it is observed that the total isoprene emission rate was highest at RZ2 (YBP) followed by RZ3 (ABP) and then RZ1 (DU). It was clearly pointed out that RZ2 had high D. sissoo population as compared to RZ1 & RZ3. Therefore, total isoprene emission rate per area in terms of this species was quite high and the isoprene concentration decreases as the number of D. sissoo plants decreases at selected sites.

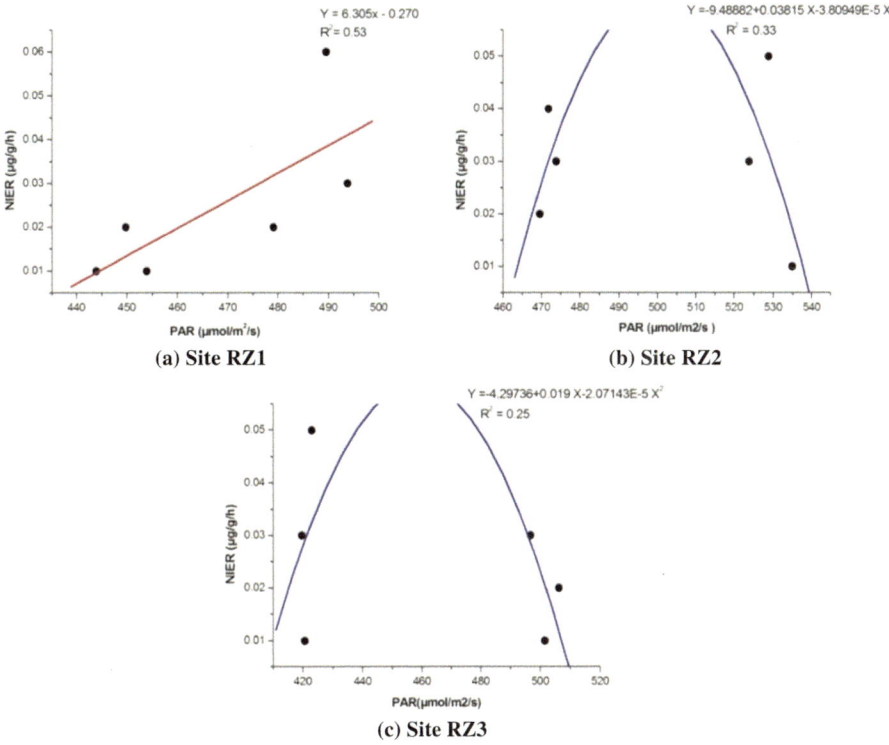

Fig. 10 (**a–c**) Variation of isoprene with PAR at different sites during winter season in *N. oleander*

4 Conclusion

The present study concludes that Dalbergia sissoo comes under high isoprene emission category, while Nerium oleander comes under BDL isoprene emission category. In addition to that, Site RZ2 (YBP) has got high isoprene emission rates as compared to Site RZ1(DU) and RZ3(ABP) which clearly depicts that in area where the population of tree species are of older category, they emit high isoprene.

As for dependence on temperature and solar radiation, isoprene emission rate of Dalbergia sissoo increases with increase in these two factors and they are positively correlated also. During summer season, high isoprene emission rates were found followed by winter and monsoon. Moreover, the significant feature of this study is the estimation of total isoprene emission rate at selected sites which shows that

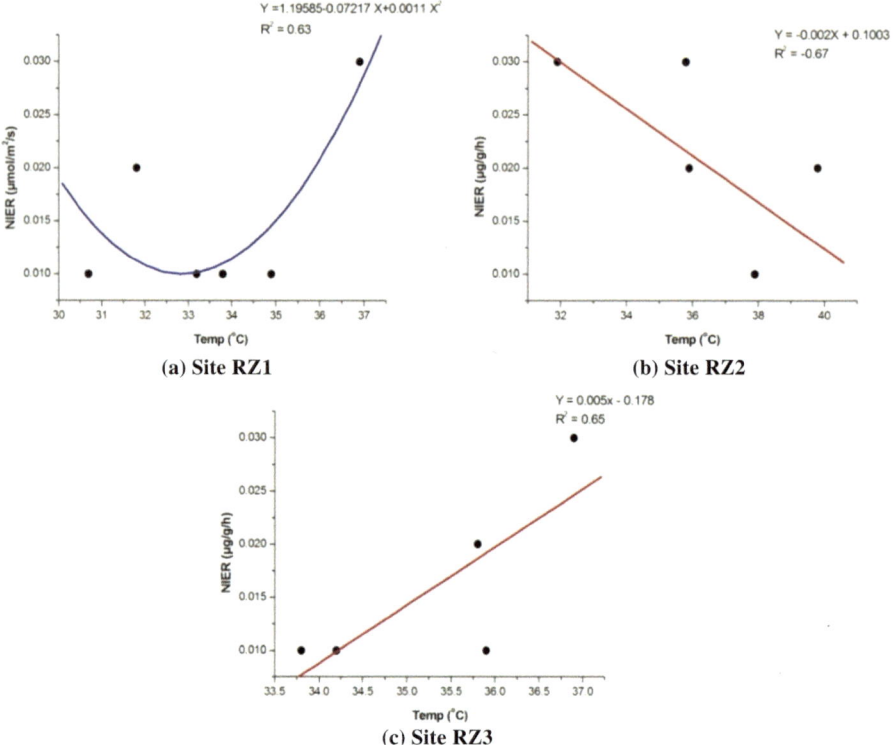

Fig. 11 (**a–c**) Variation of isoprene with enclosure temperature at different sites during monsoon season in *N. oleander*

higher the number of high isoprene emitting plant species, the higher the isoprene pollution in that area which is ultimately responsible for high production of tropospheric ozone. So, in other words, areas where D. sissoo plants are more, there are chances of high ozone concentrations for which isoprene is responsible.

For any city planner, it is very important to select the type of plant species to be planted. The present small study reflects that Nerium oleander should be planted at outskirts of selected areas and planting of Dalbergia sissoo should be done on a low scale so that the air remains clean and indirect production of tropospheric ozone, aerosol production will be minimized.

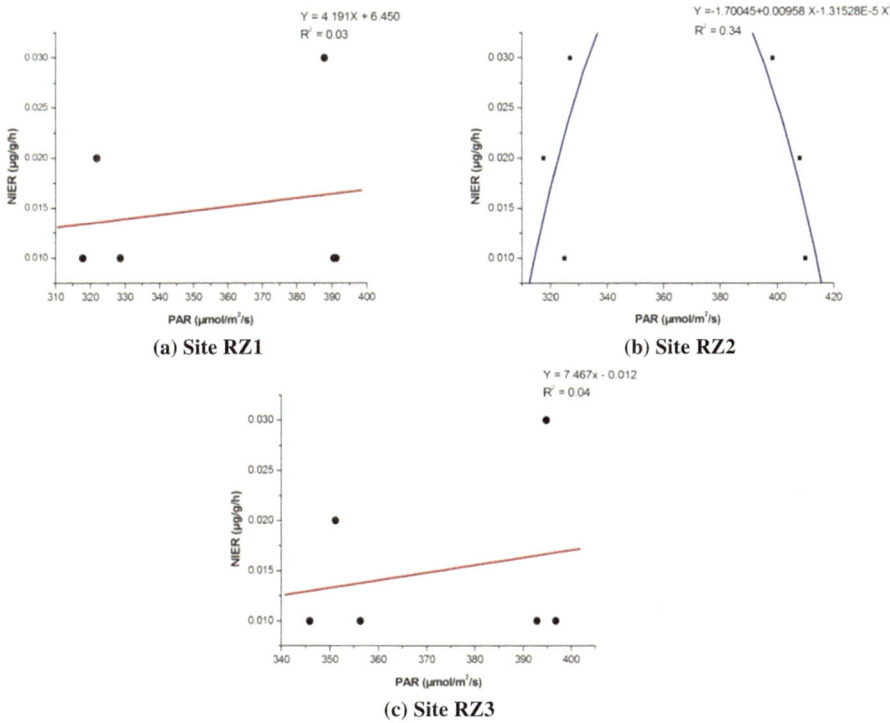

Fig. 12 (**a–c**) Variation of isoprene with PAR at different sites during monsoon season in *N. oleander*

Table 1 Diversity of selected plant species

S.N.	Sites	*D. sissoo*	*N. oleander*
1.	RZ1	50 (approx.)	30 (approx.)
2.	RZ2	3500 (approx.)	50 (approx.)
3.	RZ3	5000 (approx.)	20 (approx.)

Source: DDA, DU Fact Sheet, YBP and Fact Sheet, ABP

Table 2 Mean temperature and PAR at selected sites

S.N	Sites	Avg. Temp.	Avg. PAR	N	SD
1	RZ1	28.87	453.09	21	3.80
2	RZ2	29.34	442.90	21	2.67
3	RZ3	30.11	421.57	21	3.02

Where *Avg.* average, *N* number, *PAR* photosynthetic active radiation, *SD* standard deviation

Table 3 Mean isoprene emission rates (µg/g/h), temperature (°C), and PAR (µmol/m²/s) of (a) *Dalbergia sissoo* (Family: Fabaceae) and (b) *Nerium oleander* (Family: Apocynaceae) inside experimental setup

S.N	Sites	NIER	SD	N	Avg. Temp.	Avg. PAR
(a)						
1	RZ1	60.14	1.80	21	32.33	574.09
2	RZ2	51.65	0.77	21	36.73	586.90
3	RZ3	51.60	0.43	21	34.19	575.57
(b)						
1	RZ1	0.02	0.01	21	33.37	578.75
2	RZ2	0.03	0.01	21	37.61	592.73
3	RZ3	0.02	0.01	21	35.30	581.11

Where *Avg.* average, *NIER* Normalization of isoprene emission rate, *N* number, *PAR* photosynthetic active radiation, *SD* standard deviation

Table 4 Mean foliar mass (g dry.wt/branch) of (a) *Dalbergia sissoo* (Family: Fabaceae) and (b) *Nerium oleander* (Family: Apocynaceae) at selected sites during different seasons

Seasons	RZ1	RZ2	RZ3
(a)			
Summer	36.63 ± 2.00[a]	40.4 ± 3.08[a]	34.13 ± 1.50[a]
Winter	20.25 ± 5.02[b]	20 ± 7.21[b]	20 ± 1.55[b]
Monsoon	34.4 ± 2.12[a]	37.7 ± 1.27[a]	36.65 ± 0.63[a]
(b)			
Summer	41.6 ± 1.13[a]	44.23 ± 0.58[a]	44.57 ± 1.52[a]
Winter	26.3 ± 3.39[b]	27.35 ± 3.46[b]	24.85 ± 2.90[b]
Monsoon	43.45 ± 0.63[a]	43 ± 2.82[a]	41.9 ± 1.41[a]

Note: In the above Table 4 (a, b), each value represents mean of 6 replicates ± standard deviation. Data followed by different letters in a column are significantly different at $P \leq 0.05$. Data followed by same letters in a row are nonsignificant at $P \leq 0.05$
[a] represents the readings for Dalbergia sissoo
[b] represents the readings for Nerium oleander

Acknowledgement The authors pay sincere thanks to Dr. Ajay, Senior Technical Assistant, Advanced Instrumentation Research Facility (AIRF), Jawaharlal Nehru University (JNU), New Delhi for analysis of isoprene samples in GC-FID. The authors are also highly grateful to Council for Scientific and Industrial Research (CSIR) for awarding a Senior Research Fellowship (SRF).

References

Benjamin MT, Sudol M, Bloch L, Winer AM (1996) Low emitting urban forests: a taxonomic methodology for assigning isoprene and monoterpene emission rate. Atmos Environ 30:1437–1452
Bezemer TM, van Dam NM (2005) Linking aboveground and belowground interactions via induced plant defences. Trends Ecol Evol 20:617–624

Brilli F, Barta C, Fortunati A, Lerdau M, Loreto F, Centritto M (2007) Response of isoprene emission and carbon metabolism to drought in white poplar (Populus alba) saplings. New Phytol 175:244–254

Chameides WL, Lindsay RW, Richardson J, Kiang CS (1988) The role of biogenic hydrocarbons in urban photochemical smog: Atlanta as a case study. Science 241:1473–1475

Chen F, Tholl D, Bohlmann J, Pichersky E (2011) The family of terpene synthases in plants: a mid-size family of genes for specialized metabolism that is highly diversified throughout the kingdom. Plant J 66:212–229

Dudareva N, Andersson S, Orlova I, Gatto N, Reichelt M, Rhodes D (2005) From the cover: the nonmevalonate pathway supports both monoterpene and sesquiterpene formation in snapdragon flowers. Proc Natl Acad Sci USA 102:933–938

Estévez JM, Cantero A, Reindl A, Reichler S, Leon P (2001) 1-Deoxy-D-xylulose-5-phosphate synthase, a limiting enzyme for plastidic isoprenoid biosynthesis in plants. J Biol Chem 276:22901–22909

Fares S, Barta C, Brilli F, Centritto M, Ederli L, Ferranti F, Pasqualini S, Reale L, Tricoli D, Loreto F (2006) Impact of high ozone on isoprene emission, photosynthesis and histology of developing Populus alba leaves directly or indirectly exposed to the pollutant. Physiol Plant 128:456–465

Font X, Artola A, Sánchez A (2011) Detection, composition and treatment of volatile organic compounds from waste treatment plants. Sensors 11:4043–4059

Goyal A, Khaliq L (2011) Pulmonary functions and ambient air pollution in residents of Delhi. Ind J Med Spec 2(2):96–100

Guenther A, Hewitt CN, Erickson D, Fall R, Geron C, Graedel T (1995) A global model of natural volatile organic compound emissions. J Geophys Res 100:8873–8892

Guenther A, Karl T, Harley P, Wiedinmyer C, Palmer PI, Geron C (2006) Estimates of global terrestrial isoprene emissions using MEGAN (model of emissions of gases and aerosols from nature). Atmos Chem Phys 6:3181–3210

Harley PC, Litvak ME, Sharkey TD, Monson RK (1994) Isoprene emission from velvet bean leaves. Interactions among nitrogen availability, growth photon flux density, and leaf development. Plant Physiol 105:279–285

Harley PC, Monson RK, Lerdau MT (1999) Ecological and evolutionary aspects of isoprene emission from plants. Oecologia 118:109–123

Harley P, Vasconcellos P, Vierling L, Pinheiro CCS, Greenberg J, Guenther A, Klinger L, Almeida SS, Neill D, Baker T, Phillips O, Malhi Y (2004) Variation in potential for isoprene emissions among neotropical forest sites. Glob Chang Biol 10:630–650

Hassiotou F, Evans JR, Ludwig M, Veneklaas EJ (2009) Stomatal crypts may facilitate diffusion of CO_2 to adaxial mesophyll cells in thick sclerophylls. Plant Cell Environ 32:1596–1611

Hewitt CN (2009) Biogenic volatile organic compounds in the earth system. New Phytol 183:27–51

Iriti M, Faoro R (2009) Chemical diversity and defense metabolism: how plants cope with pathogens and ozone pollution. Int J Mol Sci 10:3371–3399

Karlik JF, Winer AM (2001) Measured isoprene emission rates of plants in California landscapes: comparison to estimates from taxonomic relationships. Atmos Environ 35:1123–1131

Knoppel H, Versino B, Peil A, Schauenburg H, Vissers H (1981) Quantitative determination of terpenes emitted by conifers. In: Proceedings of the 2nd European symposium on physicochemical behaviour of atmospheric pollutants, Varese, Italy, September 29 October 2001, Joint Research Center, Ispra, 89–98

Köksal M, Zimmer I, Schnitzler J-P, Christianson DW (2010) Structure of isoprene synthase illuminates the chemical mechanism of teragram atmospheric carbon emission. J Mol Biol 402:363–373

Lakušić B, Popov V, Runjajić-Antić D (2007) Morphoanatomical characteristics of the raw materials of the herbal drug Olivae folium and its counterfeits. Arch Biol Sci Belg 59:187–192

Loreto F, Schnitzler JP (2010) Abiotic stresses and induced BVOCs: a review. Trends Plant Sci 15:154–166

Loreto F, Forster A, Durr M, Cisky O, Seufert G (1998) On the monoterpene emission under heat stress and on the increased thermotolerance of leaves of Quercus ilex L. fumigated with selected monoterpenes. Plant Cell Environ 21:101–107

Mithofer A, Boland W (2012) Plant Defense against herbivores: chemical aspects. Annu Rev Plant Biol 63:431–450

Monson RK, Jaeger CH, Adams WW, Driggers EM, Silver GM, Fall R (1992) Relationships among isoprene emission rate, photosynthesis and isoprene synthase activity as influenced by temperature. Plant Physiol 98:1175–1180

Monson RK, Jones RT, Rosenstiel TN, Schnitzler JP (2012) Why only some plants emit isoprene. Plant Cell Environ. https://doi.org/10.1111/pce.12015

Owen SM, Peñuelas J (2005) Opportunistic emissions of volatile isoprenoids. Trends Plant Sci 10:420–426

Owen S, Boissard C, Street R, Duckham S, Csky O, Hewitt CN (1997) Screening of 18 Mediterranean plant species for VOC emissions. Atmos Environ 31:101–118

Peñuelas J, Llusia J (2003) BVOCs: plant defense against climate warming? Trends Plant Sci 8(3):105–109

Pichersky E, Sharkey TD, Gershenzon J (2006) Plant volatiles: a lack of function or a lack of knowledge? Trends Plant Sci 11:421–421

Possell M, Nicholas Hewitt C, Beerling DJ (2005) The effects of glacial atmospheric CO_2 concentrations and climate on isoprene emissions by vascular plants. Glob Chang Biol 11:60–69

Rasmussen RA (1972) What do hydrocarbons from trees contribute to air pollution. J Air Waste Manag Assoc 22:537–542

Rodriguez-Concepción M (2010) Supply of precursors for carotenoid biosynthesis in plants. Arch Biochem Biophys 504(SI):118–122

Roth-Nebelsick A, Hassiotou F, Veneklaas EJ (2009) Stomatal crypts have small effects on transpiration: a numerical model analysis. Plant Physiol 151:2018–2027

Sanadze GA, Kursanov AL (1966) On certain conditions of the evolution of the diene C5H8 from poplar leaves. Sov Plant Physiol 13:184–189

Sanadze GA, Kalandaze AN (1966) Light and temperature curves of the evolution of C5H8. Sov Plant Physiol 13:458–461

Saxena P, Ghosh C (2013) Ornamental plants as sinks and bioindicators. Environ Technol 34(21-24):3059–3067. https://doi.org/10.1080/09593330.2013.800590

Schnitzler JP, Arenz R, Steinbrecher R, Lehning A (1996) Characterization of an isoprene synthase from leaves of Quercus petracea Liebl. Bot Acta 109:216–221

Schwender J, Zeidler J, Groner R, Muller C, Focke M, Braun S (1997) Incorporation of 1-deoxy-D-xylulose into isoprene and phytol by higher plants and algae. FEBS Lett 414:129–134

SFR. State of Forest Report, Forest survey of India (2005) Government of India: ministry of environment and forest. 1–212

Sharkey TD, Singsaas EL, Vanderveer PJ, Geron C (1996) Field measurements of isoprene emission from trees in response to temperature and light. Tree Physio 16(7):649–654

Sharkey T (1996) Emission of low molecular mass hydrocarbons from plants. Trends Plant Sci 1:78–82

Sharkey TD, Singsaas EL (1995) Why plants emit isoprene? Nature 374:769

Sharkey TD, Loreto F, Delwiche CF (1991) High carbon dioxide and sun/shade effects on isoprene emission from oak and aspen tree leaves. Plant Cell Environ 14(3):333–338

Sharkey TD, Yeh SS (2001) Isoprene emission from plants. Annu Rev Plant Physiol Plant Mol Biol 52:407–436

Sharkey TD, Yeh S, Wiberley AE, Falbel TG, Gong D, Fernandez DE (2005) Evolution of the isoprene biosynthetic pathway in kudzu. Plant Physiol 137:700–712

Sharkey TD, Wiberley AE, Donohue AR (2008) Isoprene emission from plants: why and how? Ann Bot 101:5–18

Silver GM, Fall R (1991) Enzymatic synthesis of isoprene from dimethylallyl diphosphate in aspen leaf extracts. Plant Physiol 97:1588–1591

Singsaas EL, Sharkey TD (2000) The effects of high temperature on isoprene synthesis in oak leaves. Plant Cell Environ 23(7):751–757

Singsaas EL, Lerdau M, Winter K, Sharkey TD (1997) Isoprene increases thermotolerance of isoprene-emitting leaves. Plant Physiol 115:1413–1420

Singsaas EL, Laporte MM, Shi JZ, Monson RK, Bowling DR, Johnson K, Lerdau MK, Jasentuliytana A, Sharkey TD (1999) Leaf temperature fluctuation affects isoprene emission from red oak (Quercus rubra L.) leaves. Tree Physiol 19:917–924

Street RA, Duckham S, Hewitt CN (1996) Laboratory and field studies of biogenic VOC emissions from Sitka spruce *(Picea sitchensisbong)* in the UK. J Geophys Res 101:22799–22806

Tao Z, Jain AK (2005) Modeling of global biogenic emissions for key indirect greenhouse gases and their response to atmospheric CO_2 increases and changes in land cover and climate. J Geophys Res 110:D213–D209

Tingey DT, Manning M, Grothaus LC, Burns WF (1979) The influence of light and temperature on isoprene emission rates from live oak. Physiol Plant 47:112–118

USEPA (Centre for Environmental Research Information) (1999) TO-15, Compendium of Methods for the Determination of Volatile Organic Compounds (VOCs) in Air Collected in Specially-Prepared Canisters and Analyzed by Gas Chromatography/Mass Spectrometry (GC/MS)

Westberg H, Lamb B, Kempf K, Allwine G (2000) Isoprene emission inventory for the BOREAS southern study area. Tree Physiol 20:735–743

Zimmerman PR, Chatfield RB, Fishman J, Crutzen PJ, Hanst PL (1978) Estimation of the production of CO_2 and H_2 from the oxidation of hydrocarbon emission from vegetation. Geophys Res Lett 5:679–682

Microplastics: An Unsafe Pathway from Aquatic Environment to Health—A Review

Guncha Sharma and Chirashree Ghosh

Abstract Waste is considered to be any solid material that is manufactured or processed and then discarded or disposed, ending up either in a terrestrial or an aquatic environment. It is possible to curb the negative effects of waste when it is dumped on land, but as it enters water, its potential impact is unpredictable, as water is a universal solvent with an ability to dissolve and hoard contaminants. Contaminants can be organic and inorganic, which determine their associated toxicity. Various solid wastes such as plastic, glass, metal, rubber, and wood have already polluted available potable water resources. Of all contaminants, plastics are commonly used presently and although they reduce the exploitation of natural resources, they are highly persistent in nature and end up in water bodies in the form of plastic debris. Plastic debris within an aquatic environment can be macro, meso, micro, or nano in size. Larger macro-size particles have been reported to commonly entangle aquatic invertebrates, birds, mammals, and turtles. On the other hand, micro- and nano-size particles form through different fragmentation processes such as photodegradation and microbial degradation, and as a result of other external forces. Micro-sized plastics (<5 mm in diameter) are ubiquitous in water, sediment, and their associated biota. Their smaller size and colorful appearance makes them resemble food for aquatic organisms, and hence they bioaccumulate easily. Also, additives used during plastic synthesis are released into the aquatic and terrestrial environments in the form of toxic chemicals. Among anthropogenic "cosmetics," toothpaste is considered to be one of the major contributors to microplastics pollution. It uses micro-sized beads for exfoliation. Research on microplastics is rapidly increasing worldwide, and microplastics have already been tagged as an alarming threat, but there exists a knowledge gap on actual concentrations, their impacts, and their sources. This review addresses possible types and sources of microplastics, their bioavailability, and an analysis of their fate, or potential impact, on aquatic environments.

Keywords Microplastics · Contaminants · Toxicity · Environment · Aquatic health

---

G. Sharma (✉) · C. Ghosh
Department of Environmental Studies, University of Delhi, Delhi, India

© The Author(s), under exclusive licence to Springer Nature Switzerland AG 2019
T. Jindal, *Emerging Issues in Ecology and Environmental Science*,
SpringerBriefs in Environmental Science, https://doi.org/10.1007/978-3-319-99398-0_5

1   Introduction

Earth is the only planet to sustain life because it provides vital commodities. The water, air, soil, flora, and fauna always interact with one another. If any one vanishes, the entire cycle is disturbed, as each has an important role. All available natural resources (both renewable and nonrenewable) are finite, so one cannot use them without thinking of sustainability. These days, because of rapid urbanization and changing lifestyles, resources are being overexploited, leading to their scarcity for future generations. In addition to overexploitation, toxic substances are continuously added to the environment, leading to the larger problem of pollution.

Solid waste is a reason for environmental degradation. Nowadays people prefer packaged foods and beverages and use plastic containers and tools. The use of plastic increases as the population increases. Thus a major component of discarded trash is plastic—items such as bags, bottles, food containers, and wrappers. Without concern, people choose bodies of water as an expedient place to dispose of solid plastic waste, and through different processes this plastic is converted into smaller particles.

1.1   Plastic as a Waste

Plastic is human friendly because of its vibrant colors, durability, light weight, strength, and most important, affordability. The word *plastic* derives from the Greek term *plasticos*, which means material that is malleable and can be easily shaped. Plastics are derived from the cracking of carbon-based naphtha that has been extracted from crude oil. Under strong heat and pressure, the crude oil fractures, thereby changing its molecular structure. This entire process takes place in petrochemical refineries. In general, plastic is made up of *polymers*, which are substances made up of subunits called *monomers*. These monomers join together to form patterns of long, branched chains and form different types of polymers, such as polyethylene, polypropylene, polystyrene, and polyvinyl chloride.

Plastic is a nondegradable compound, has a long half-life, and hence persists in the environment for thousands of years. After being disposed, plastic waste is burned and melted, during which numerous harmful gases and toxic substances are released; these affect the health of living organisms. Plastic waste has severe socioeconomic impacts such as downturns in tourism because of loss of the aesthetic value of the environment. The amount of plastic waste in the environment depends on various factors such as proximity to beaches, migrant population settlements, accidental spillage, and proximity to industrial or commercial areas.

Once plastic waste is directly or indirectly dumped into the environment, it ultimately enters the aquatic food chain. Then, through fragmentation processes such as microbial degradation (biofilm formation, biofouling), photo-oxidation, and other mixing factors (wind, current, abrasion, etc.), plastic debris changes from mega-sized pieces (>100 m in diameter) to macro- (>20 mm in diameter), micro- (between 5 and 20 mm in diameter), and nano-sized particles (<5 mm).

2 Microplastics

Large plastic particles have relatively fixed sizes, whereas microplastics have various sizes; these range from <1 mm (Browne et al. 2010; Claessens et al. 2011), <2 mm (Ryan et al. 2009), and < 5 mm (Barnes et al. 2009) to 2–6 mm (Derraik 2002). Because of their small size, microplastics have a high probability of transferring into the food chain. To date, sewage treatment plants, plastic pellet production industries, and the cosmetics industry are the main sources of microplastic pollution, which ends up in drainage systems and eventually water bodies. Because of their low buoyancy, plastics travels long distances over water surfaces, polluting ecosystems and accumulating in sediments.

2.1 Types of Microplastics

Microplastics are divided into two categories on the basis of origin: primary and secondary microplastics. *Primary microplastics* are plastics that are intentionally generated for commercial purposes, such as plastic resins; these are also known as "*mermaid tears.*" *Secondary microplastics* are larger plastics that break down to smaller sizes and undergo further changes through hydrolytic processes, such as the mechanical action of waves and water currents, oxidative degradation, or the photo-catalytic action of UV radiation. Fendall and Sewell (2009) indicated that some particulates and fibers are small enough to be filtered out of wastewater treatment plants, thereby creating problems.

2.2 Effects of Microplastics

Scientists have reported various effects of microplastics on organisms (Gregory 1991; Ryan et al. 2009; Eerkes-medrano et al. 2015). Ingestion of microplastic particles blocks the digestive pathway, causing choking, internal and external injuries, ulcerating sores, starvation, and gradual weakening of the body, which ultimately cause illness. Studies have reported that microplastics of a size <10 μm can translocate from the gut to the circulatory system of mussels (Browne et al. 2007; Setälä et al. 2014). Once ingested, these microplastics can be either eliminated through defecation or retained within the tissues of the exposed animals; this is called *translocation*. The accumulation of such fragments in sediment inhibits gaseous exchange between the overlying water and the water in the porous sediment, resulting in hypoxia or anoxia at the benthic level, which interferes with normal air exchange and alters organisms living on the sea floor (Goldberg 1994; Derraik 2002). In general, ingestion of microplastics leads to blockages of the intestinal tract and secretion of gastric enzymes, reduced steroid hormone levels, and delayed ovulation, which ultimately leads to reproductive failure and death (Azzarello and Van Vleet 1987; Derraik 2002). Microplastics resemble food for aquatic organisms, and after ingestion they not only accumulate in the stomach but also reduce the eating efficiency of organisms, known as *pseudo-satiation*; that is, more space in the stomach is occupied by plastic particles, and thus there is less room for actual food. Microplastics are hydrophobic, so waterborne organic pollutants adhere to them and cause toxicity in the aquatic biota. Thus ingestion of microplastics may introduce toxins to the food chain and ultimately leads to biomagnification. Figure 1 describes the microplastics cycle.

Researchers have found evidence that some seabirds select microplastic particles with specific shapes and colors, mistaking them for food and potential prey, thereby disturbing feed selectivity (Carpenter et al. 1972; Wright et al. 2013). Studies have concluded that ingestion of plastic particles directly correlates with foraging strategies, foraging techniques, and diet (Moser and Lee 1992; Derraik 2002). According to studies by Laist (1987) and Fry et al. (1987), accumulation of microplastics in adult birds could allow the microplastics to be passed to the birds' offspring during feeding. Fulmars were used as a model to monitor and estimate the abundance of plastic debris in water bodies, and the amount of plastic in stomachs of fulmars is now considered to be one marker of ecological quality (van Franeker 2010). This

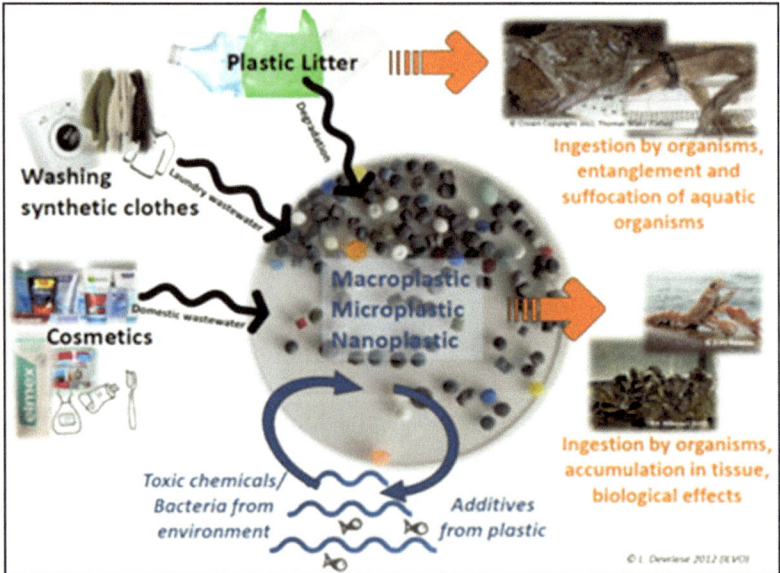

Fig. 1 Microplastic cycle (Source: Lisa Devriese, Institute for Agriculture and Fisheries Research, ILVO, ©2012)

indicates that the health of a broad range of organisms—from suspension, filter, and deposit feeders to detritivores, planktivorous fish, mammals, and birds, as well as human beings—is at risk and of concern globally.

3 Conclusion

Microplastics are widely distributed throughout the environment, and this is becoming a looming problem. On a global scale, microplastics are already considered to be an alarming threat to aquatic biota, but a huge knowledge gap still exists at the national level. To date, most studies have mainly focused on the marine environment, and few have studied freshwater bodies. Most important, microplastic concentrations needs to be assessed along various dimensions, such as their relations with hydrodynamic factors (e.g., current velocity, turbidity), anthropogenic factors (e.g., urban areas and industrial zones), meteorological factors (e.g., temperature and wind), and geographical location (e.g., slopes, beaches). The major challenge in estimating microplastic concentrations in different domains of an environment is the lack of standardized sampling and separation techniques. Future studies are needed to build the relation between physicochemical conditions and the abundance of microplastics in that particular ecosystem. There also is a dire need to understand the impacts of microplastics on biological organisms.

Acknowledgment The author acknowledges the Delhi University Library System for providing access to the literature database. The author also thanks the Delhi University Grant Commission for providing a fellowship, and the Delhi University Research Council research and development scheme.

References

Azzarello MY, Van Vleet ES (1987) Marine birds and plastic pollution. Mar Ecol Prog Ser 37:295–303. https://doi.org/10.3354/meps037295
Barnes DKA, Galgani F, Thompson RC, Barlaz M (2009) Accumulation and fragmentation of plastic debris in global environments. Philos Trans R Soc Lond B Biol Sci 364(1526):1985–1998. https://doi.org/10.1098/rstb.2008.0205
Browne MA, Galloway T, Thompson R, Chapman PM (2007) Learned discourses. Integr Environ Assess Manag 3(4):2004–2006
Browne MA, Galloway TS, Thompson RC (2010) Spatial patterns of plastic debris along estuarine shorelines. Environ Sci Technol 44(9):3404–3409. https://doi.org/10.1021/es903784e
Carpenter EJ, Anderson SJ, Harvey GR, Miklas HP, Peck BB (1972) Polystyrene spherules in coastal waters. Science 178:749–750
Claessens M, De Meester S, Van Landuyt L, De Clerck K, Janssen CR (2011) Occurrence and distribution of microplastics in marine sediments along the Belgian coast. Mar Pollut Bull 62(10):2199–2204. https://doi.org/10.1016/j.marpolbul.2011.06.030
Derraik JGB (2002) The pollution of the marine environment by plastic debris: a review. Mar Pollut Bull 44:842–852. https://doi.org/10.1016/S0025-326X(02)00220-5

Eerkes-medrano D, Thompson RC, Aldridge DC (2015) ScienceDirect microplastics in freshwater systems: a review of the emerging threats, identification of knowledge gaps and prioritisation of research needs. Water Res 75:63–82. https://doi.org/10.1016/j.watres.2015.02.012

Fendall LS, Sewell MA (2009) Contributing to marine pollution by washing your face: microplastics in facial cleansers. Mar Pollut Bull 58(8):1225–1228. https://doi.org/10.1016/j.marpolbul.2009.04.025

Fry DM, Fefer SI, Sileo L (1987) Ingestion of plastic debris by Laysan albatrosses and wedge-tailed shearwaters in the Hawaiian Islands. Mar Pollut Bull 18(6 Suppl B):339–343. https://doi.org/10.1016/S0025-326X(87)80022-X

Goldberg ED (1994) Diamonds and plastics are forever? Mar Pollut Bull 28(8):466. https://doi.org/10.1016/0025-326X(94)90511-8

Gregory MR (1991) The hazards of persistent marine pollution: drift plastics and conservation islands. J R Soc N Z 21(2):83–100. https://doi.org/10.1080/03036758.1991.10431398

Laist DW (1987) Overview of the biological effects of lost and discarded plastic debris in the marine environment. Mar Pollut Bull 18(6 Suppl B):319–326. https://doi.org/10.1016/S0025-326X(87)80019-X

Lisa Devriese (2012) Institute for Agriculture and Fisheries Research, ILVO

Moser ML, Lee DS (1992) A fourteen-year survey of plastic ingestion by western North Atlantic seabirds. Colon Waterbirds 15(1):83–94

Ryan PG, Moore CJ, van Franeker JA, Moloney CL (2009) Monitoring the abundance of plastic debris in the marine environment. Philos Trans R Soc Lond B Biol Sci 364(1526):1999–2012. https://doi.org/10.1098/rstb.2008.0207

Setälä O, Fleming-lehtinen V, Lehtiniemi M (2014) Ingestion and transfer of microplastics in the planktonic food web. Environ Pollut 185:77–83. https://doi.org/10.1016/j.envpol.2013.10.013

van Franeker JA (2010) Fulmar litter EcoQO monitoring in the Netherlands 1979–2008 in relation to EU Directive 2000/59/EC on port reception facilities. IMARES, Wageningen UR

Wright SL, Thompson RC, Galloway TS (2013) The physical impacts of microplastics on marine organisms: a review. Environ Pollut 178:483–492. https://doi.org/10.1016/j.envpol.2013.02.031

A Review on the Dairy Industry Waste Water Characteristics, Its Impact on Environment and Treatment Possibilities

Surbhi Sinha, Abhinav Srivastava, Tithi Mehrotra, and Rachana Singh

Abstract Dairy industry is one of the most polluting industries in India. Due to the elevated milk demand, the dairy industry in India has developed swiftly, leading to a large amount of waste discharge in the nearby water bodies. The waste water from dairy industry is characterized by high BOD, COD, organic and inorganic contents. Release of these waste waters into the water bodies without suitable remediation can cause serious environmental issue. Indian government has enforced very stern rules and regulations for the waste water discharge to safeguard the environment. Therefore, suitable methods are needed to meet the effluent discharge standards. This chapter thus discusses the various sources and characterization of dairy waste water, their impact on the environment and the conventional as well as the improved techniques for the treatment of dairy waste water.

Keywords Dairy industry · Sources · Characteristics · Treatment · Advanced treatment technologies

1 Introduction

Dairy industry is considered to be one of the major industries of food sector playing an important role in the economy of the country. India has attained first position in milk production, out of all the milk-producing nations and is sharing about 13.1% of the total milk produced in the world (Kumbhar 2010). Dairy industry is believed to have a notable effect on the water pollution as there are about 286 large- and small-scale dairy industries in India accountable for plenty of waste production, both in solid and liquid form (Kothari et al. 2012). Approximately, 110 million tons of milk and 275 million tons of waste water are released annually from the indian dairy industries (Kushwaha et al. 2011). Expeditious expansion of dairy industries has not only increased the efficiency of work rate but has also resulted in the production and discharge of dangerous stuff into the environment, consequently

S. Sinha · A. Srivastava · T. Mehrotra · R. Singh (✉)
Amity Institute of Biotechnology, Amity University, Noida, Uttar Pradesh, India
e-mail: ssinha2@amity.edu; rsingh2@amity.edu

© The Author(s), under exclusive licence to Springer Nature Switzerland AG 2019
T. Jindal, *Emerging Issues in Ecology and Environmental Science*,
SpringerBriefs in Environmental Science, https://doi.org/10.1007/978-3-319-99398-0_6

Table 1 Minimal standards for discharge of effluent from dairy industry (Bharati and Shinkar 2013a, b)

Parameters	World Bank Report	CPCB, India
pH	6–9	6.5–8.5
BOD$_5$	50	100 (Based on BOD$_1$)
COD	250	–
TSS	50	150
Oil and grease	10	10
Total nitrogen	10	–
Total phosphorous	2	–
Temperature increase	≤3 °C	–

Except pH, all parameters are in terms of mg L^{-1}

causing health hazards and affecting flora and fauna. The dairy industry is one of those sectors where the cleaning tanks, homogenizers, pipe sand, heat exchangers and other equipment release a huge amount of effluents with a high organic load. This organic load is basically constituted of milk (raw material and dairy products), reflecting effluent with high levels of chemical oxygen demand (COD), biochemical oxygen demand (BOD), oil and grease, nitrogen and phosphorus (Srivastava et al. 2016). This demands rapid and efficient treatment of the waste water before being discharged into the environment. Water management in dairy industry is well recognized, but production and disposal of effluent remains a challenging subject. The Indian government has enforced very stern rules and regulations for the discharge of effluent in order to protect the environment (Table 1). Accordingly, suitable techniques should be used by the industries to meet the effluent discharge standards.

To facilitate dairy industry to contribute in water conservation, development of a well-organized and cost-effective technique is essential. In order to have proper processes in the effluent treatment plant, characterization of waste water, treatability studies and planning of proper units and processes for effluent treatment is very much necessary.

2 Sources of Dairy Waste Water

Effluent from the dairy industry emerges from the following parts of the plant: receiving station, cheese plant, bottling plant, casein plant, butter plant, dried milk plant, condensed milk plant and ice cream plant. Waste from the dairy is also generated from the washing and cleaning out of the product remaining in the cans, trucks, pipes or other equipments. Spillage of the products due to overflows, leaks, careless handling, boiling over also are sources of the dairy waste. Moreover, sludge discharge from settling tank or discharges from the bottles and detergents used in washing also lead to the origin of dairy waste.

Table 2 Characteristics of dairy industry waste water

Waste type	COD	BOD	pH	TSS	TS	References
Dairy effluent	1900–2700	1200–1800	7.2–8.8	500–740	900–1350	Sukhadev et al. (2013)
Milk and dairy products factory	10,251.2	4840.6	8.34	5802.6	–	Cristian (2010)
Arab dairy factory	3383 ± 1345	1941 ± 864	7.9 ± 1.2	831 ± 392		Tawfik et al. (2008)
Whey	71,526	20,000	4.1	22,050	56,782	Deshpande et al. (2012)
Cheese whey pressed	80,000–90,000	1,20,000–1,35,000	6	8000–11,000		Tikariha and Sahu (2014)
Aavin dairy industry wash water	2500–3300	–	6.4–7.1	630–730	1300–1400	Sathyamoorthy and Saseetharan (2012)

Except pH, all parameters are in terms of mg L^{-1}

3   Characteristic of the Dairy Effluent

The effluent from dairy industry contains plenty of milk constituents such as casein, inorganic salts besides sanitizers and detergents used for washing. It also has dissolved sugars, fats, proteins and possibly residues of additives. Dairy waste water is characterized by intense smell, high COD, BOD, dust and variable pH (Kothari et al. 2011). The waste load equivalent of specific milk constituents are as follows: 1 kg of milk fat = 3 kg of COD; 1 kg of lactose = 1.13 kg of COD; 1 kg of protein = 1.36 kg of COD (Kasmi et al. 2017). It contains adequate amount of nutrients like nitrogen (14–830 mg L^{-1}) and phosphorous (9–280 mg L^{-1}) which can promote the growth of pathogens (Rico Gutierrez et al. 1991). Additionally, it has dissolved solids, suspended solids, chlorides, sulphates, oil, and grease. Typically, the characteristics of dairy effluent largely depend on the quantity of milk processed and the product manufactured. Table 2 shows the typical characteristics of dairy effluent reported by various authors.

4   Effects of Dairy Effluent on Environment

Waste generated from dairy industry contains highly putrecible organic constituents (Qasim and Mane 2013). The organic constituents present in the dairy waste water putrefy fast, diminishing the level of dissolved oxygen in the receiving water bodies creating anaerobic conditions and intense foul odour. Also, the receiving water becomes breeding place for mosquitoes and flies which can cause threatening diseases like malaria, chikungunya, and dengue (Kumar and Desai 2011). The dairy effluent contains loads of milk components like inorganic salts, casein along with detergents and sanitizers used for washing. All these constituents are basically responsible for the increase in high BOD and COD (Singh et al. 2014) of the water, which surpasses the permissible limit set by Bureau of Indian Standard (BIS). The casein precipitation from the dairy waste decomposes further into odorous black sludge. Additionally, dairy wastes are also hazardous to certain fish. These effluents promote the growth of certain algae and bacteria which eventually utilize oxygen present in the water, consequently leading to the death of fish due to suffocation. Dairy industries result in the emission of toxic gases like carbon dioxide, sulphur oxides and nitrogen oxide to the atmosphere. Furthermore, methane is released during anaerobic treatment and nitrous oxide from the soil at dairy waste waster irrigation sites. Carbon dioxide, methane and nitrous oxide are greenhouse gases, and their increased emission can lead to global warming. Global warming in turn would affect the environment in many ways including increased melting of snow and ice, rise in sea level, and desertification. Apart from global warming, these greenhouse gases also lead to ocean acidification, ozone depletion, smog pollution as well as changes to plant growth and nutrition. Dairy industries also emit particulate material in the atmosphere leading to increase in dust which settles down on the surrounding building, making them vulnerable to corrosion.

5 Dairy Waste Water Treatment

Treatment methods generally used in dairy industry are the following.

5.1 Mechanical Treatment

This is the initial phase of the dairy waste treatment and involves screens, grit chambers, skimming tank or sedimentation tank. The screens are in the inclined position with the direction of the flow. The large materials floating in the water bodies or effluent are removed by screens; or else they can choke up the small pipes which can further affect the working of effluent pumps. Grit chambers can remove the heavier inorganic particles such as grit and sand. Skimming tanks are used to remove oil, grease, fruit skins, wood pieces, etc. A sedimentation tank allows suspended particles to settle out of waste water as it flows slowly through the tank. Sludge is formed at the bottom of the tank which is further again treated to make it less toxic (Patel et al. 2016).

5.2 Chemical Treatment

Chemical treatment involves the technique of precipitation. The method requires the addition the flocculants like aluminium salt, iron salt, and lime to the waste water and strenuous mixing using agitators. This leads to the precipitation of inorganic phosphate in the form of very fine particles which then combine together to form larger flocks. These flocks then sit down at the bottom of the sedimentation tank as primary sludge, whereas the clear effluent passes into the basin for biological treatment (Birwal et al. 2017).

5.3 Biological Treatment

Biological treatment often referred to as secondary treatment is used to remove materials left after primary treatment. It is considered to be one of the dependable methods for the dairy effluent treatment. Since dairy waste water mainly comprises organic waste, biological treatment method is the favourable choice for the removal of these organic wastes as these methods utilize soluble compounds and small colloids. Based on oxygen requirements, biological treatments are of two main types: aerobic and anaerobic (Sengil and Ozacar 2006).

In aerobic biological treatment system, the micro-organisms in the presence of oxygen oxidize organics to carbon dioxide and water, whereas anaerobic system is generally used for the treatment of high strength dairy waste water where micro-organisms in the absence of oxygen convert organic matter and nutrients to methane and CO_2, while the rest of biomass is used for cell growth and maintenance. Table 3 compares the various advantages and disadvantages of aerobic and anaerobic treatment system for dairy waste water.

These days generally aerobic systems are used for the treatment of dairy effluent, but there are number of drawbacks associated with these processes. High energy requirement, filamentous growth and rapid acidification due to high lactose level and low water buffer capacity are the some of the downsides related with aerobic systems. Anaerobic systems however are more reliable for the treatment of dairy waste water loaded with a high organic content. This process of treatment is a cost-effective process as it does not require aeration or large area and generates vey less amount of sludge.

Table 3 Comparison of advantages and disadvantages of aerobic and anaerobic treatment of dairy industry waste water (Bharati and Shinkar 2013a, b)

Factors	Aerobic process	Anaerobic process
Reactors	Aerated lagoons, stabilized ponds, trickling filters, biological discs	UASB, anaerobic filter upflow packed bed reactor, CSTR downflow fixed film reactor
Reactor size	Large size	Smaller reactor size
Energy	High energy is required	These processes produce energy in the form of methane
Biomass yield	6–8 times greater biomass is produced as compared to anaerobic process	Lower biomass is produced
Shock loading	Excellent performance in this regard	Not very good response to shock loading

6   Advanced Technologies for Dairy Effluent Treatment

6.1   *Physicochemical Process*

Electrocoagulation

Electrocoagulation is one of the advanced alternative processes for the treatment of dairy waste water. It is carried out by employing the electric current across the metal plates immersed in water. Organic and inorganic wastes, heavy metals, colloidal particles, etc. are primarily held in water by electrical charges. By applying another electrical charge to the contaminated water, the charges that hold the particles together are destabilized and separate from the clean water. The particles then coagulate to form a mass, which can be easily removed. This process is a low sludge producing technique. Above all, waste waters treated by electrocoagulation are clear, palatable, colourless and odourless. However, the use of electricity makes it an expensive process. Sharma (2014) removed COD and oil from dairy waste water using electrocoagulation. The batch experimental results showed 87% of COD removal at 3 A, pH 9 and electrolysis time of 75 min. Shivayogimath and Vijayalaxmi (2014) used electrocoagulation technique for the dairy waste water treatment. During the study they found that 98.75% COD and 97.82% turbidity can be removed with very short electrolysis duration of 10 min at applied voltage of 7 V and pH 6. Gerson de Freitas Silva et al. (2015) studied the efficiency of electrocoagulation using aluminium electrodes in treating waste water from a dairy industry. They reported 57% COD removal, 99% turbidity removal and 92% removal of total suspended solids at pH 5 and current density of 61.6 A m^{-2} (Fig. 1).

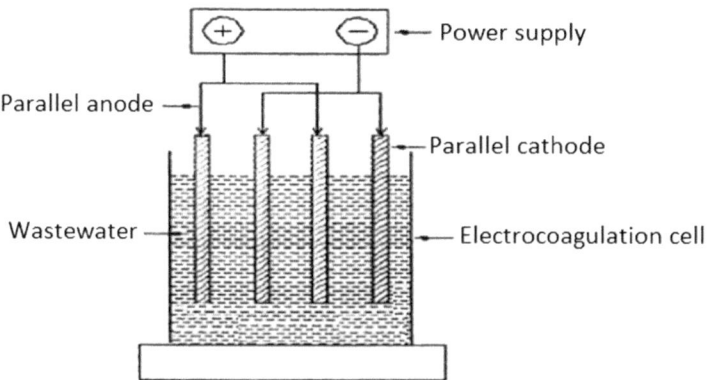

Fig. 1 Schematic design of electrocoagulation Reactor. Source: Dehghani et al. (2015)

Adsorption

Adsorption is a waste water treatment technique used for eliminating large number of non-degradable organic compounds from waste water. Activated carbon is the most widely used adsorbent for the treatment of waste water. Some low cost adsorbents like rice husk, coal fly ash and straw dust are also used for the treatment of waste water. Pathak et al. (2016) used rice husk as an adsorbent for the removal of organic pollutants from dairy effluent. Maximum removal of as high as 92.5% could be achieved using an adsorbent dosage 5 g L^{-1}, pH 2, and temperature 30 °C. Marol et al. (2017) treated dairy waste water using sugar bagasse ash and rice husk as adsorbents. Kanawade and Bhusal (2015) used activated charcoal for the treatment of dairy waste water. Activated charcoal removed a maximum of 65% COD and 67% BOD from dairy effluent.

Membrane Treatment

Membrane separation plays an imperative role in waste water treatment, water reclamation and desalination applications. The common membrane separation processes like microfiltration, ultrafiltration, nanofiltration, reverse osmosis and electrodialysis are used to remove contaminants from dairy waste water. These processes function effectively at low temperatures and low energy requirements. Above all, high feasible product recovery is possible. However, equipment costs are high and fouling of membrane takes place which can lead to decrease in the flux of permeates. Zielińska and Galik (2017) removed approximately 90% of COD from dairy waste water using micro filteration system. A high-performance bioreactor, an aerobic jet loop reactor, combined with a ceramic membrane filtration unit, was used to investigate its suitability for the treatment of the dairy processing waste water. A loading rate of 53 kg COD m^{-3} d^{-1} resulted in 97–98% COD removal efficiencies fewer than 3 h hydraulic retention time (Burhanettin and Suleyman 2011). Andrade et al. (2015) evaluated the technical and economic feasibility of membrane bioreactors followed by nanofiltration for dairy waste water treatment in order to reuse the treated effluent. It was observed that the membrane bioreactors efficiently removed the organic matter and colour of the feed effluent (Fig. 2).

6.2 Aerobic Treatment

Sequencing Batch Reactor (SBR)

SBRs are one of the most promising types of activated sludge treatment where the entire treatment process takes place in a set of tanks that usually run on a fill and draw basis. The tanks may be a rectangular basin, an earthen or oxidation ditch, or any other concrete/metal type structure. This technique treats the waste water in

Fig. 2 Simple schematic describing the membrane bioreactor process. Source: Le-Clech et al. (2006)

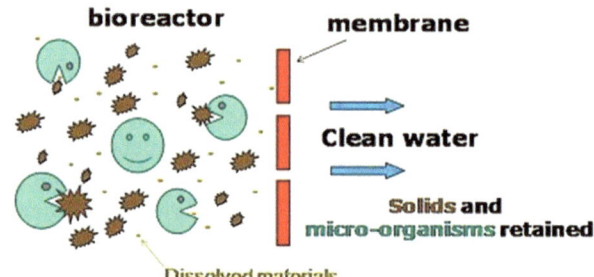

batch mode and each batch is sequenced through a series of treatment stages. Each tank in the SBR is filled with the waste water and mixed with biomass. Aeration is provided to the tank to encourage biological growth and waste reduction. After a discrete period of time, both mixing and aeration are stopped and the solids are allowed to settle down at the bottom of the tank while clarified effluent is decanted. This process is simple and the quality of effluent after treatment is high. Moreover, it reduces main pollutants like ammonia and phosphate. However, sludge must be disposed off frequently and it sometimes becomes difficult to adjust cycle times for small communities. Mohseni-Bandpi and Bazari (2004) investigated a bench-scale aerobic SBR to treat the waste water from an industrial milk factory. The results showed 90% of COD removal in all conditions. Li and Zhang (2002) investigated the performances of single-stage and two-stage sequencing batch reactor (SBR) for treating dairy waste water. It was found that two-stage system consisting of an SBR and a complete-mix biofilm reactor was capable of achieving complete ammonia oxidation and comparable carbon, solids, and nitrogen removal while using at least 1/3 less HRT as compared to the single SBR system (Fig. 3).

6.3 Anaerobic Treatment

Upflow Anaerobic Sludge Blanket (UASB)

Upflow anaerobic sludge blanket reactor is one of the most widely used anaerobic digesters for the treatment of industrial waste water. It is a single tank process consisting of anaerobic micro-organisms for the treatment of industrial waste water, resulting in almost complete removal of organic pollutants. Waste water infiltrates the reactor from the bottom and then flows upward. A suspended sludge blanket in the reactor treats the waste water as it flows through it. Micro-organisms in the sludge break down the organic matter present in the waste water by anaerobic digestion into methane and carbon dioxide. Sludge production is low as compared to aerobic treatment system due to the slow growth rate of anaerobic organisms and good removal efficiency is achieved even at high loading rates and low temperatures. Howbeit, pathogens are only partially removed in the UASB reactor and

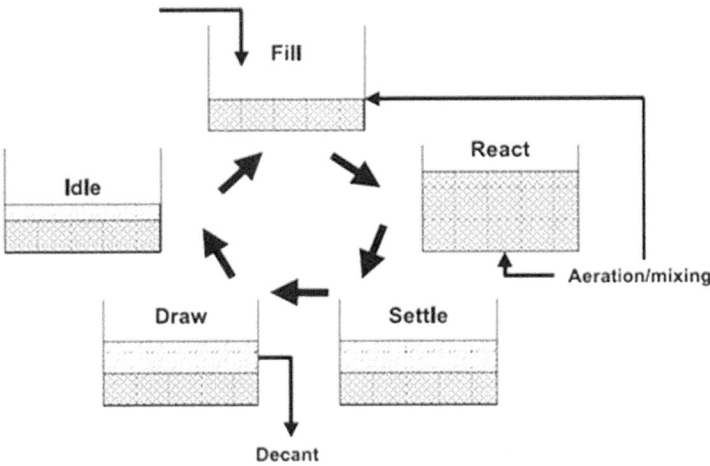

Fig. 3 Typical cycles of SBR. Source: USEPA (1999)

production of hydrogen sulphide during the treatment can lead to bad smell and corrosion. Gavala et al. (1999) treated dairy waste water using an upflow anaerobic sludge blanket reactor. They found that operation at an organic loading rate of 6.2 g COD L^{-1} d was found to be safe and could be increased to a maximum of 7.5 g COD L^{-1} d and COD reduction of 90% at organic loading rate of 0.031 kg COD m^{-3} d was achieved operating in steady-state conditions using a waste water with a COD influent of 2050 mg L^{-1}. UASB reactor was used for the treatment of dairy waste water by Kavitha et al. (2013). COD and BOD removal of 77% and 87%, respectively was achieved in the reactor (Fig. 4).

Anaerobic Sequencing Batch Reactors (ASBR)

ASBR is a newly developed batch reactor system that includes digestion and separation of solids within a single vessel. ASBR follows four steps for the treatment of waste water: feeding, reaction, settling and withdrawal of treated effluent. These reactors are extensively used because of their merits like these are simple, have efficient quality control of effluents, diminishes the settling step for both affluent and effluent and also can be used to treat a variety of effluents. However, one major limitation with ASBR is that they do not perform well when overloaded. Abdulsalam et al. (2011) treated synthetic milk waste water in anaerobic sequence batch reactor. The percentage COD reduction at 35 °C with additional seeds has been observed at the lower organic loading 1 g L^{-1} and retention time of 72 h to be 83.33%. In one lab-scale study, thermophilic ASBR and mesophilic ASBR systems provided volatile organic solid removal between 26% and 44%, and between 26% and 50% for dairy waste water respectively (Nadais et al. 2005). In another lab-scale study, 62% and 75% removal rate of COD and BOD was observed at a hydraulic retention

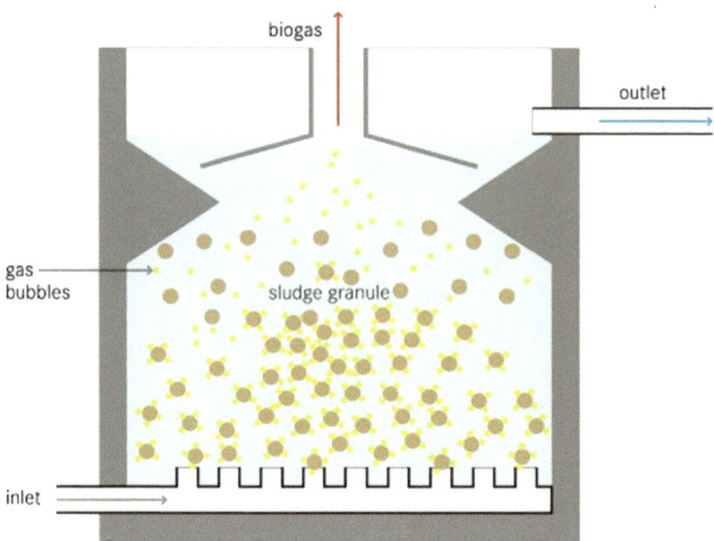

Fig. 4 Schematic diagram of an upflow anaerobic sludge blanket reactor (UASB). Source: Tilley et al. (2014)

time (HRT) of 6 h, at 58 °C, for a substrate containing non-fat dry milk (Dugba and Zhang 1999).

7 Conclusion

Dairy waste water causes serious water pollution and harms the entire ecosystem, and its treatment therefore becomes a primary need for industries. Industries can use techniques that are inexpensive and can be operated with minimum supervision. Biological methods are considered to be most effective for the treatment of dairy waste water, where aerobic systems are easier to operate and control and anaerobic systems produce less amount of sludge with lower energy requirement. Since none of the techniques alone is capable of treating the dairy waste water efficiently, a combined process can be developed specifically to treat the dairy waste water complying with the minimum effluent discharge requirement.

References

Abdulsalam TD, Arinjay K, Sambi SS (2011) Study on anaerobic treatment of synthetic milk wastewater under variable experimental conditions. Int J Environ Sci Develop 2:17–23
Andrade LH, Mendes FDS, Espindola JC, Amaral MCS (2015) Reuse of dairy wastewater treated by membrane bioreactor and nanofiltration: technical and economic feasibility. Braz J Chem Eng 32:735–747. https://doi.org/10.1590/0104-6632.20150323s00003133

Bharati SS, Shinkar NN (2013a) Dairy industry wastewater sources, characteristics & its effects on environment. Int J Curr Eng Technol 3:1611–1615

Bharati SS, Shinkar NP (2013b) Comparative study of various treatments for dairy industry wastewater. IOSR J Eng 3:42–47. https://doi.org/10.9790/3021-03844247

Birwal P, Deshmukh G, Priyanka, Saurabh SP (2017) Advanced technologies for dairy effluent treatment. J Food Nutr Popul Health 1:1–5

Burhanettin F, Suleyman U (2011) The investigation of dairy industry wastewater treatment in a biological high performance membrane system. Biochem Eng J 57:46–54. https://doi.org/10.1016/j.bej.2011.08.007

Cristian O (2010) Characteristics of the untreated wastewater produced by food industry. Analele Universităţii din Oradea, Fascicula:Protecţia Mediului 15:709–714

Dehghani M, Sheibani SS, Taghizadeh MM (2015) Optimization of organic compounds removal from wastewater by electrocoagulation. J Hormo Univ Med Sci 19:59–65

Deshpande DP, Patil PJ, Anekar SV (2012) Biomethanation of dairy waste. Res J Chem Sci 2:35–39

Dugba P, Zhang R (1999) Treatment of dairy wastewater with two stage anaerobic sequencing batch reactor systems: thermophilic versus mesophilic operations. Bioresour Technol 68:225–233. https://doi.org/10.1016/S0960-8524(98)00156-4

Gavala HN, Kopsinis H, Skiadas IV, Stamatelatou K, Lyberatos G (1999) Treatment of dairy wastewater using an upflow anaerobic sludge blanket reactor. J Agric Eng Res 73:59–63. https://doi.org/10.1006/jaer.1998.039

Gerson de Freitas Silva V, Regina Célia Santos M, José Antônio Marques P (2015) The efficiency of electrocoagulation using aluminum electrodesin treating wastewater from a dairy industry. Cienc Rural 45:1713–1719. https://doi.org/10.1080/03601234.2012.646174

Kanawade SM, Bhusal VC (2015) Adsorption on dairy industrial wastewater by using activated charcoal as adsorbent. Int J Chem Mater Sci 3:025–032

Kasmi M, Hamdi M, Trabelsi I (2017) Processed milk waste recycling via thermal pretreatment and lactic acid bacteria fermentation. Environ Sci Pollut Res 24:13604. https://doi.org/10.1007/s11356-017-8932-6

Kavitha RV, Shiva K, Suresh R, Krishnamurthy V (2013) Performance evaluation and biological treatment of dairy waste water treatment plant by upflow anaerobic sludge blanket reactor. Int J Chem Petrochem Technol 3:9–20

Kothari R, Virendra K, Tyagi VV (2011) Assessment of waste treatment and energy recovery from dairy industrial waste by anaerobic digestion. Inst Integ Omic Appl Biotechnol 2:1–6

Kothari R, Vinayak VP, Kumar V, Singh DP (2012) Experimental study for growth potential of unicellular alga Chlorella pyrenoidosa on dairywaste water: an integrated approach for treatment and biofuel production. Bioresour Technol 116:466–470. https://doi.org/10.1016/j.biortech.2012.03.121

Kumar D, Desai K (2011) Pollution abatement in milk and dairy industry. Curr Pharma Res 1:45–152

Kumbhar V (2010) Livestock sector in India–recent trends and progress. International Business. Article source, Business http://EzineArticles.com/?expert=vijay_kumbhar

Kushwaha JP, Srivastava VC, Mall ID (2011) An overview of various technologies for the treatment of dairy wastewaters. Crit Rev Food Sci 51:442–452. https://doi.org/10.1080/10408391003663879

Le-Clech P, Chen V, Fane AG (2006) Fouling in membrane bioreactors used in wastewater treatment. J Membr Sci 284:17–53. https://doi.org/10.1016/j.memsci.2006.08.019

Li X, Zhang R (2002) Aerobic treatment of dairy wastewater with sequencing batch reactor systems. Bioprocess Biosyst Eng 25:103. https://doi.org/10.1007/s00449-002-0286-9

Marol C, Seema S, Biradar S, Chavan S, Badiger S (2017) Treatment of dairy industry wastewater by adsorption method. Int J Adv Eng Res Develop 4:505–507

Mohseni-Bandpi A, Bazari H (2004) Biological treatment of dairy wastewater by sequencing batch reactor. Iranian J Environ Health Sci Eng 1:65–69

Nadais H, Capela I, Arroja L, Duarte A (2005) Optimum cycle time for intermittent UASB reactors treating dairy wastewater. Water Res 39:1511–1518. https://doi.org/10.1016/j.watres.2005.01.020

Patel A, Sharma S, Mitra S, Shah M (2016) Performance and evaluation study of dairy wastewater. Int J Adv Technol Eng Sci 4:172–176

Pathak U, Das P, Banerjee P, Datta S (2016) Treatment of wastewater from a dairy industry using Rice husk as adsorbent: treatment efficiency, isotherm, thermodynamics, and kinetics modelling. J Thermodyn 2016:1–7. https://doi.org/10.1155/2016/3746316

Qasim W, Mane AV (2013) Characterization and treatment of selected food industrial effluents by coagulation and adsorption techniques. Water Resour Ind 4:1–12. https://doi.org/10.1016/j.wri.2013.09.005

Rico Gutierrez JL, Garcia Encina PA, Fdz-Polanco F (1991) Anaerobic treatment of cheese-production wastewater using a UASB reactor. Bioresour Technol 37:271–276. https://doi.org/10.1016/0960-8524(91)90194-O

Sathyamoorthy GL, Saseetharan MK (2012) Dairy wastewater treatment by anaerobic hybrid reactor—a study on the reactor performance and optimum percentage of inert media fill inside reactor. Res J Chem Environ 16:51–56

Sengil A, Ozacar M (2006) Treatment of dairy wastewaters by electrocoagulation using mild steel electrodes. J Hazard Mater 137:1197–1205. https://doi.org/10.1016/j.jhazmat.2006.04.009

Sharma D (2014) Treatment of dairy waste water by electro coagulation using aluminum electrodes and settling, filtration studies. Int J Chem Tech Res 6:591–599

Shivayogimath CB, Vijayalaxmi RN (2014) Treatment of dairy industry wastewater using electro-coagulation technique. Int J Eng Res Technol 3

Singh NB, Singh R, Imam MM (2014) Waste water management in dairy industry: pollution abatement and preventive attitudes. Int J Sci Environ Technol 3:672–683

Srivastava AK, Rana SVS, Mehrortra T, Singh R (2016) Characterization and immobilization of bacterial consortium for its application in degradation of dairy effluent. J Pure Appl Microbiol 10:2199–2208

Sukhadev VS, Kulkarni SW, Wani M (2013) Physicochemical characterization of dairy effluents. Int J Life Sci Biotechnol Pharm Res 2:182–191

Tawfik A, Sobheyb M, Badawya M (2008) Treatment of a combined dairy and domestic wastewater in an up-flow anaerobic sludge blanket (UASB) reactor followed by activated sludge (AS system). Desalination 227:167–177. https://doi.org/10.1016/j.desal.2007.06.023

Tikariha A, Sahu O (2014) Study of characteristics and treatments of dairy industry waste water. J. Appl Environ Microbiol 2:16–22. https://doi.org/10.12691/jaem-2-1-4

Tilley E, Ulrich L, Lüthi C, Reymond P, Zurbrügg C (2014) Compendium of sanitation systems and technologies, 2nd revised edn. Swiss Federal Institute of Aquatic Science and Technology (Eawag), Duebendorf

United States Environmental Protection Agency (USEPA) (1999) Wastewater technology fact sheet: sequencing batch reactors. U.S. Environmental Protection Agency, Office of Water, Washington, DC EPA 932-F-99-073

Zielińska M, Galik M (2017) Use of ceramic membranes in a membrane filtration supported by coagulation for the treatment of dairy wastewater. Water Air Soil Pollut 228:173. https://doi.org/10.1007/s11270-017-3365-x

Physicochemical Treatment of Research Laboratory Wastewater: A Case Study

Sundar Ramanathan and R. B. Lal

Abstract Research laboratory professionals explore the wonder of science through experiments that generate new knowledge and sometimes, new products. In India, there are many research laboratories in government sector as well as in private sector. Each research laboratory's diverse developmental activities generate comparatively small quantities of wastes with widely varying compositions. These wastes may include acidic/toxic fumes, wastewater contaminated with oils, heavy metals and organics, discarded solid residues, and contaminated solid containers. These wastes are released into the environment through sink disposal of chemicals, release of chemicals through fume hoods, and mixing of laboratory wastes with other trash material.

Until the 1980s research laboratory wastes were not a great concern. Legal and regulatory requirements, reinforced by a public opinion spurred the handling of hazardous wastes in a responsible way. With the stricter environmental legislation, there can be civil and criminal penalties for failure to legal requirements. So it was necessary to develop and improve the ways of managing wastes generated in the laboratory.

In developed countries, research laboratory wastes are managed more professionally and in an effective manner. In India, management of research laboratory wastes was receiving due attention, only recently, and there was a need for in depth R&D in this area. The quantity and quality of the research laboratory wastes are varied from laboratory to laboratory, and it mainly depends upon the type of research orientation, experimental work, consumption of analytical and chemical reagents and the amount of experimental wastes. Direct disposal of such laboratory wastes to the environment might lead to a serious surface water pollution as well as ground water pollution. So there should be a suitable treatment methodology to handle such type of wastes. As a result, there was a growing interest in the development of new technologies and procedures, for the decontamination of the research laboratory wastes.

Note: The content of this chapter is the views of authors only and not that of the organization.

S. Ramanathan (✉)
Ministry of Environment, Forests and Climate Change, Regional Office, Chennai, India

R.B.Lal
Ministry of Environment, Forests and Climate Change, New Delhi, India

T. Jindal, *Emerging Issues in Ecology and Environmental Science*,
SpringerBriefs in Environmental Science, https://doi.org/10.1007/978-3-319-99398-0_7

In this scenario, one can consider the various treatments options, which includes physicochemical treatment and biological treatment. It was essential to characterize the wastes in terms of physical, chemical, and biochemical parameters before adopting a treatment sequence. The case study presents the detailed waste management studies carried out for a research laboratory, which involved in research and development work in the field of environmental science. It includes the evaluation of the quantity and quality of liquid waste generation and to delineate proper treatment sequence to reduce the pollution load from the laboratory wastewater for reuse (or) safe disposal into surface water/public sewer/irrigation purpose without causing any pollution problem.

Keywords Research lab · Waste water · Effluent · Chemical treatment

1 Introduction

Research laboratory professionals explore the wonder of science through experiments that generate new knowledge and sometimes, new products. In India, there are many research laboratories in Government sector as well as in private sector. Each research laboratory's diverse developmental activities generate comparatively small quantities of wastes with widely varying compositions. These wastes may include acidic/toxic fumes, wastewater contaminated with oils, heavy metals and organics, discarded solid residues, and contaminated solid containers. These wastes are released in to the environment through sink disposal of chemicals, release of chemicals through fume hoods, and mixing of laboratory wastes with other trash material.

Until the 1980s research laboratory wastes were not a great concern. Legal and regulatory requirements, reinforced by a public opinion spurred the handling of hazardous wastes in a responsible way. With the stricter environmental legislation, there can be civil and criminal penalties for failure to legal requirements. So it was necessary to develop and improve the ways of managing wastes generated in the laboratory.

In developed countries, research laboratory wastes are managed more professionally and in an effective manner. In India, management of research laboratory wastes has received due attention only recently, and there is a need for in-depth R&D in this area. The quantity and quality of the research laboratory wastes are varied from laboratory to laboratory, and it mainly depends upon the type of research orientation, experimental work, consumption of analytical and chemical reagents, and the amount of experimental wastes. Direct disposal of such laboratory wastes to the environment might lead to a serious surface water pollution as well as ground water pollution. So there should be a suitable treatment methodology to handle such type of wastes. As a result, there was a growing interest in the development of new technologies and procedures, for the decontamination of the research laboratory wastes.

In this scenario, one can consider the various treatments options, which includes physicochemical treatment and biological treatment. It was essential to characterize the wastes in terms of physical, chemical, and biochemical parameters before adopting a treatment sequence. This case study presents the detailed waste management studies carried out for a research laboratory, which involved in research and development work in the field of environmental science. It includes the evaluation of the quantity and quality of liquid waste generation and to delineate proper treatment sequence to reduce the pollution load from the laboratory wastewater for reuse (or) safe disposal in to surface water/public sewer/irrigation purpose without causing any pollution problem.

2   About the Research Laboratory Under Study

The case study was conducted during the period of October 2004 to June 2005. The research laboratory was spread over nearly about 108 acres. Among that 40% of the total land area was under forest cover. The research laboratories were mainly handling lot of R&D activities, which includes sponsored R&D projects, developmental projects and In-house research projects in the field of environmental science. From the field survey, the water supply and wastewater conveyance systems of the research laboratories were studied.

3   Water Supply System and Wastewater Generation

The water supply to the laboratory premises was accomplished through two municipal water supply meters. Every day from the overhead reservoir, approximately 80 m^3 water was supplied to laboratory tanks by means of gravity flow. Water was mainly consumed by various types of experimental work, chemical analysis, glassware cleaning, floor washing, drinking, and lavatory purposes.

The water consumed by the research laboratories are finally getting discharged in a manhole which further conveys to municipal drainage systems. The direct discharge method was used to determine the laboratory wastewater outflow. These flow measurements were taken on hourly basis. The details of flow measurements are shown in Table 1.

From the details of the flow measurements, found that the average flow rate of wastewater from the research laboratories was around 40–50 m^3 everyday.

4   Quality of the Wastewater

A detailed chemical inventorization study was carried out in that research laboratory, to infer the quality of the wastewater through the consumption of analytical and chemical reagents. The chemicals inventorization study indicated that, the mixture of organic and inorganic chemicals, heavy metals and organic solvents might contaminate the laboratory wastewater. This study also revealed that, proper treatment was needed before discharging such type of wastewater in to public sewer/irrigation purpose/inland surface water. The list of major solvents consumed by the laboratories for the past 6 years was shown in Table 2.

Table 1 Flow measurement details

Set. No.	Average flow rate of wastewater (L/day)	Peak factor	Set. No.	Average flowrate of wastewater (L/day)	Peak factor
1.	44,226 ± 8845.12	1.121	7.	43,079 ± 502	1.151
2.	37,638 ± 4817	1.318	8.	43,503 ± 956.39	1.140
3.	39,096 ± 4033.76	1.269	9.	43,892 ± 824.12	1.130
4.	49,615 ± 4015.2	1.000	10.	44,120 ± 932.20	1.124
5.	44,170.92 ± 1688.5	1.123	11.	43,526 ± 630.20	1.139
6.	42,768 ± 981.70	1.160	12.	44,520 ± 836	1.114

Average daily flow rate of wastewater = 43,346.16 ± 2421.81 L/day

Table 2 List of major solvents and their consumption

S. No.	Name of solvents	Quantity (L)
1.	Acetone	265
2.	Acetonitrile	188
3.	Benzene	132.5
4.	Chloroform	108.5
5.	Cyclohexane	74.5
6.	Diethylether	37
7.	Methanol	178.5
8.	n-Hexane	111.5
Total		1095.5

5 Sampling

The effluent samples were collected from that specified manhole on hourly basis. The method of sampling was grab sampling and composite sampling. According to the standard methods, samples were collected from the specified manhole and the characterization was done in terms of physical, chemical, and biochemical parameters. The details of instruments used and measurement methods adopted for the characterization of the laboratory wastewater are as follows (Table 3).

6 Results and Discussion

6.1 Characteristics of Laboratory Wastewater

The BOD value of the wastewater was founded to exceed the permissible limit of 30 mg/L as per the Indian standards for Industrial and sewage effluents discharge IS: 2490-1982. The lead content of the wastewater was founded to exceed the IS: 2490-1982. Contamination with phenolic compounds, cyanides was also observed.

Table 3 Instruments used/measurement methods adopted

S. No.	Parameters	Measurement method/Instrument used
1.	pH	Electrometric method/pH meter
2.	Colour	'HACH' DR2010 Portable data logging spectrophotometer
3.	Total Dissolved solids	TDS Filtration assembly/Oven
4.	Chemical oxygen Demand	Open reflux method/Hot plate digestion
5.	Biochemical oxygen Demand	5 day B.O.D. test/D.O. (Winkler method), Incubator
6.	Total organic carbon	High Temperature Combustion method/'Thermos Electron Corporation' T.O.C 1200 Analyzer
7.	Phenols	Direct photometric method/"Bausch and Lomb" spectronic-21D Spectrophotometer (500 nm wavelength)
8.	Cyanides	Titrimetric method
9.	Heavy metals	"PerkinElmer" JV –24 ICP—AES, "PerkinElmer" optima 4100DV ICP—OES/"Mile stone" ETHIOS Microwave digestion system/Flame photometer
10.	Mercury	"Perkin Elmer" Mercury Analyzer
11.	Jar test Apparatus (Treatability studies)	'Phipps and bird' Multiple stirring device Jar tester (American Society for Testing and Materials 2000)
12.	Toxic Characteristic Leaching Procedure	Zero headspace extraction vessels (USEPA 1986)
13.	BET Surface area measurement (Activated carbon)	"Micromeritrics Instrument Corporation" 2010K Accelerated Surface Area Porosimetry
14.	Carbon, Hydrogen, Sulfur, Nitrogen measurement (Activated carbon)	"Carlo Elba" 1108 CHNS-O Elemental analyzer

The characterization results revealed that proper treatment was needed to discharge the wastewater into public sewer/irrigation land/inland surface water bodies. Characteristics of the research laboratory wastewater were given in Tables 4 and 5. The average analysis results (five sets of analysis) of grab samples and composite samples are represented in Table 4. Then it was compared with Indian standards (IS: 2490-1982) for the discharge of industrial and sewage effluents in to public sewer/ irrigation land/inland surface water in Table 5 (Table 6).

6.2 Methodology of Treatment

The results of analysis of the laboratory wastewater show that BOD/COD ratio of the wastewater was in the range of 0.192 ± 0.072. The concentration of heavy metals and other parameters of the wastewater are widely varying in nature. And also

Table 4 Results of analysis of grab and composite samples

Parameters	Units	Grab samples Set No. 1–5 Nos.					Average	Composite samples Set No 1–5 Nos.					Average
Colour	Pt-Co	49	54	42	29	43.5	43.5 ± 9.39	32	36	40	38	45	38.2 ± 4.816
pH	–	7.49	7.35	7.38	7.46	7.53	7.44 ± 0.07	7.51	7.62	7.67	7.56	7.54	7.58 ± 0.06
TDS	mg/L	1228	1121	1581	1612	1708	1450 ± 258.5	1321	1541	1589	1546	1684	1536.2 ± 133.288
Total Hardness	mg/L	320	310	308	342	373	330.6 ± 27.27	310	363	334	316	332	331 ± 20.61
COD	mg/L	112.8	238	81.93	91.42	44.70	113.77 ± 73.68	163.2	172.8	202.4	211.2	199.13	189.746 ± 20.61
BOD	mg/L	25	26.66	26.66	27.16	17.33	24.56 ± 4.12	40	35	40	20	35	34 ± 8.21
BOD/COD	–	0.227	0.270	0.404	0.372	0378	0.330 ± 0.072	0.295	0.202	0.197	0.091	0.175	0.192 ± 0.072
TOC	mg/L	42.30	87.38	31.23	33.54	15.52	41.99 ± 27.14	61.30	63.80	74.32	79.23	73.42	70.41 ± 0.028
COD/TOC	–	2.66	2.720	2.623	2.71	2.88	2.71 ± 0.090	2.66	2.708	2.723	2.665	2.712	2.69 ± 0.02
Phenols	mg/L	BDL	BDL	BDL	BDL	0.120	0.120 ± 0	BDL	BDL	BDL	BDL	BDL	BDL
Cyanides	mg/L	BDL	BDL	BDL	BDL	0.100	0.100 ± 0	BDL	BDL	BDL	BDL	BDL	BDL
Mercury	mg/L	0.003	0.0015	0.005	0.007	0.0055	0.0044 ± 0.002	0.006	0.002	0.0012	0.002	0.004	0.003 ± 0.001
Arsenic	mg/L	0.142	0.131	0.149	0.155	0.146	0.1446 ± 0.008	0.011	BDL	0.132	0.141	0.112	0.079 ± 0.068
Zinc	mg/L	0.423	0.061	0.272	0.075	0.139	0.194 ± 0.152	0.088	0.083	0.365	0.188	0.423	0.229 ± 0.157
Lead	mg/L	0.111	0.523	0.313	0.366	0.304	0.323 ± 0.147	0.038	0.021	BDL	BDL	0.031	0.018 ± 0.017
Nickel	mg/L	0.8620	1.120	1.253	BDL	1.253	0.8976 ± 0.52	0.003	0.004	BDL	BDL	0.004	0.0022 ± 0.017
Cadmium	mg/L	0.038	0.011	0.0687	0.041	0.055	0.0427 ± 0.02	0.022	0.011	BDL	BDL	0.033	0.0132 ± 0.014
Cobalt	mg/L	0.0131	0.0232	0.0412	BDL	0.0412	0.0265 ± 0.02	BDL	BDL	BDL	BDL	BDL	BDL
Manganese	mg/L	0.112	BDL	BDL	0.244	0.133	0.097 ± 0.102	0.068	0.044	BDL	BDL	0.012	0.024 ± 0.030
Iron	mg/L	BDL	0.689	0.363	BDL	0.425	0.295 ± 0.296	0.184	0.276	0.160	0.199	0.283	0.180 ± 0.114
Total Chromium	mg/L	0.066	0.012	BDL	0.077	0.096	0.0502 ± 0.041	0.082	0.035	0.380	BDL	0.038	0.146 ± 0.024
Copper	mg/L	BDL	0.054	BDL	0.063	0.050	0.0334 ± 0.030	0.059	0.028	BDL	BDL	0.016	0.020 ± 0.024
Aluminium	mg/L	0.051	BDL	BDL	0.492	0.811	0.2708 ± 0.365	BDL	BDL	0.596	0.692	0.413	0.340 ± 0.326

BDL below detection limit

Table 5 Results of analysis of grab and composite samples

| Parameters | Units | Results of analysis of grab and composite samples | | Indian standards for Industrial and sewage effluent discharge IS: 2490-1982 | | |
		Grab samples Average	Composite samples Average	Inland surface	Public sewer	Land for irrigation
pH	–	43.5 ± 9.39	38.2 ± 4.816	5.5–9.0	5.5–9.0	5.5–9.0
Colour	Pt-Co	7.44 ± 0.07	7.58 ± 0.06	NS	NS	NS
Total dissolved solids	mg/L	1450 ± 258.5	1536.2 ± 133.288	2100	2100	NS
Total Hardness	mg/L	330.6 ± 27.27	331 ± 20.61	NS	NS	NS
C.O.D.	mg/L	113.77 ± 73.68	189.746 ± 20.61	250	NS	NS
B.O.D.	mg/L	24.56 ± 4.12	34 ± 8.21	30	350	100
B.O.D./C.O.D.	–	0.330 ± 0.072	0.192 ± 0.072	NS	NS	NS
T.O.C.	mg/L	41.99 ± 27.14	70.41 ± 0.028	NS	NS	NS
C.O.D./T.O.C.	–	2.71 ± 0.090	2.69 ± 0.02	NS	NS	NS
Phenols	mg/L	0.120 ± 0	BDL	1.0	5.0	NS
Cyanides	mg/L	0.100 ± 0	BDL	0.2	0.2	0.2

BDL below detection limit, *NS* not stipulated

the quality of the wastewater depends upon the type of research orientation, experimental work, consumption of analytical and chemical reagents and the amount of experimental wastes. Hence the biological treatment was not feasible for this wastewater because of the BOD–COD ratio was too low.

In this scenario, physicochemical wastewater treatment was thought to be a suitable option for this research laboratory wastewater. This treatment requires smaller land area, lower initial capital costs, insensitivity to toxic materials such as heavy metals, and ability to cope with overload situations.

The physicochemical treatment, i.e., chemical coagulation by the addition of coagulants with the help of coagulant aids for the removal of heavy metals and dissolved solids from the wastewater, was evaluated. In addition, investigations were also done to assess the performance of activated carbon for the removal of residual COD from the laboratory wastewater effluent after treatment with coagulants. The sludge obtained from the chemical coagulation treatment was characterized as per the standard methods.

Table 6 Results of analysis of grab and composite samples

		Results of analysis of grab and composite samples		Indian standards for industrial and sewage effluent discharge IS: 2490-1982		
Parameters	Units	Grab samples Average	Composite samples Average	Inland surface	Public sewer	Land for irrigation
Mercury	mg/L	0.0044 ± 0.002	0.003 ± 0.001	0.01	0.01	NS
Arsenic	mg/L	0.1446 ± 0.008	0.079 ± 0.068	0.2	0.2	0.2
Zinc	mg/L	0.194 ± 0.152	0.229 ± 0.157	5.0	15	NS
Lead	mg/L	0.323 ± 0.147	0.018 ± 0.017	0.1	1.0	NS
Nickel	mg/L	0.8976 ± 0.52	0.0022 ± 0.017	3.0	3.0	NS
Cadmium	mg/L	0.0427 ± 0.02	0.0132 ± 0.014	2.0	1.0	NS
Cobalt	mg/L	0.0265 ± 0.02	BDL	NS	NS	NS
Manganese	mg/L	0.097 ± 0.102	0.024 ± 0.030	2.0	2.0	NS
Iron	mg/L	0.295 ± 0.296	0.180 ± 0.114	3.0	3.0	NS
Total Chromium	mg/L	0.0502 ± 0.041	0.146 ± 0.024	2.0	2.0	NS
Copper	mg/L	0.0334 ± 0.030	0.020 ± 0.024	3.0	3.0	NS
Aluminium	mg/L	0.2708 ± 0.365	0.340 ± 0.326	NS	NS	NS

BDL below detection limit, *NS* not stipulated

6.3   Treatability Studies

Evaluation of different coagulants with coagulant aids for the removal of heavy metals and dissolved solids from the laboratory wastewater.

The chemical treatability studies of the influent wastewater were done by using conventional coagulants, viz., lime and alum. Along with that commercially available cationic polyelectrolyte was used as a coagulant aid. From the laboratory studies, the minimum dosage for lime, alum, lime plus alum, and lime plus cationic polyelectrolyte was found to be 300 mg/L, 240 mg/L, 300 + 50 mg/L, and 300 + 1.0 mg/L, respectively. Then as per the standard methods, a series of jar test experiments were conducted by varying the dosage of each coagulant from its minimum required range. From the jar test experiments, the optimum dosage for lime, alum, lime plus alum, and lime plus cationic polyelectrolyte was found to be 500 mg/L, 350 mg/L, 400 + 150 mg/L, and 400 + 2.0 mg/L, respectively. The optimum pH for lime, alum, lime plus alum, and lime plus cationic polyelectrolyte was found to be 12.90, 7.02, 7.63, and 11.70, respectively. From the characterization studies, it was observed that there was a maximum removal of heavy metals and COD content occurred when using the lime and cationic polyelectrolyte combination at a dosage and pH of 400 + 2.0 mg/L and 11.70 respectively.

The characteristics of lime and cationic polyelectrolyte treated wastewater were given in Table 7.

Table 7 Results of analysis of coagulant-treated wastewater

Parameters	Units	Influent (Raw water)	Effluent (Coagulant treated wastewater)	Inland surface	Public sewer	Land for irrigation
		Results of analysis of coagulant-treated wastewater		Indian standards for industrial and sewage effluent discharge IS: 2490-1982		
pH	–	38.2 ± 4.816	11.70 ± 0.1136	5.5–9.0	5.5–9.0	5.5–9.0
Colour	Pt-Co	7.58 ± 0.06	BDL	NS	NS	NS
Total dissolved solids	mg/L	1536.2 ± 133.288	416 ± 5.42	2100	2100	NS
Total Hardness	mg/L	331 ± 20.61	470 ± 7.16	NS	NS	NS
C.O.D.	mg/L	189.746 ± 20.61	120.72 ± 13.97	250	NS	NS
B.O.D.	mg/L	34 ± 8.21	17 ± 5.70	30	350	100
B.O.D./C.O.D.	–	0.192 ± 0.072	0.143 ± 0.050	NS	NS	NS
Zinc	mg/L	0.229 ± 0.157	0.0022 ± 0.004	5.0	15.0	NS
Lead	mg/L	0.018 ± 0.017	BDL	0.1	1.0	NS
Nickel	mg/L	0.0022 ± 0.017	0.0011 ± 0.002	3.0	3.0	NS
Cadmium	mg/L	0.0132 ± 0.014	BDL	2.0	1.0	NS
Cobalt	mg/L	BDL	0.0052 ± 0.011	NS	NS	NS
Manganese	mg/L	0.024 ± 0.030	BDL	2.0	2.0	NS
Ferrous	mg/L	0.180 ± 0.114	BDL	3.0	3.0	NS
Total Chromium	mg/L	0.146 ± 0.024	0.0164 ± 0.028	2.0	2.0	NS
Copper	mg/L	0.020 ± 0.024	0.001 ± 0.002	3.0	3.0	NS
Aluminium	mg/L	0.340 ± 0.326	BDL	NS	NS	NS
Mercury	mg/L	0.003	BDL	0.01	0.01	NS

BDL below detection limit, *NS* not stipulated

6.4 Evaluation of the Effectiveness of Activated Carbon for the Removal of Residual COD from the Laboratory Wastewater After Chemical Coagulation

Since the wastewater after chemical coagulation, had generated an effluent having residual COD and dissolved solid content in the range of 120.72 ± 13.97 mg/L and 416 ± 5.42 mg/L respectively. The characterization results revealed that further reduction in residual COD and dissolved solids content was necessary before its reuse or disposal. Thus after chemical coagulation, the effluent was subjected to activated carbon adsorption treatment.

Table 8 Physical properties of activated carbon

Physical properties	Units	Commercial activated carbon
Total surface area (N_2 BET method)	m^2/g	190.56
Micro pore volume	cm^3/g	0.07620
Adsorption average pore diameter	$A°$	16.8245
Ash content	%	4.44
Moisture content	%	9.250
Carbon	%	122.7988
Hydrogen	%	0
Nitrogen	%	9.8658
Sulfur	%	0
Molecular formula	–	$C_{14}H_0N_1S_0$

Adsorbent Specifications

The efficiency of adsorption on activated carbon was directly related to its granular size and carbon content. Accordingly, the activated carbon used in the present study was subjected to characterization in terms of its general proximate and ultimate characteristics. The activated carbon used for this study was a commercial one. The average particle size of the activated carbon was around 1.5 mm. The adsorbent used in this study was washed in de-ionized water before contacting with the adsorbate solutions. The physical properties of the activated carbon are given in Table 8.

Bench-Scale Experimental Methods

The fixed bed column study was carried out for the estimation of the breakthrough curve for the commercial activated carbon. The coagulant-treated effluent was fed to a feed tank, having a volume of 20 L and from which effluent was pumped using a peristaltic pump at a constant flow rate of 10 mL/min. The effluent solution was fed through a bed of commercial activated carbon in up flow mode.

The carbon beds were contained in glass columns with a diameter of 30 mm and standard bed height of 300 mm. The carbon bed in the column was supported by a layer of glass wool. Approximately 100 g of activated carbon was fed in to the column and samples were drawn at regular time interval from the column outlet. The collected samples were analyzed for COD parameter.

Breakthrough Curves

The experimental results obtained from the bench-scale adsorber are illustrated as a breakthrough curve in Fig. 1. The results indicated that the activated carbon column system was successful in removing the residual COD from the research laboratory wastewater.

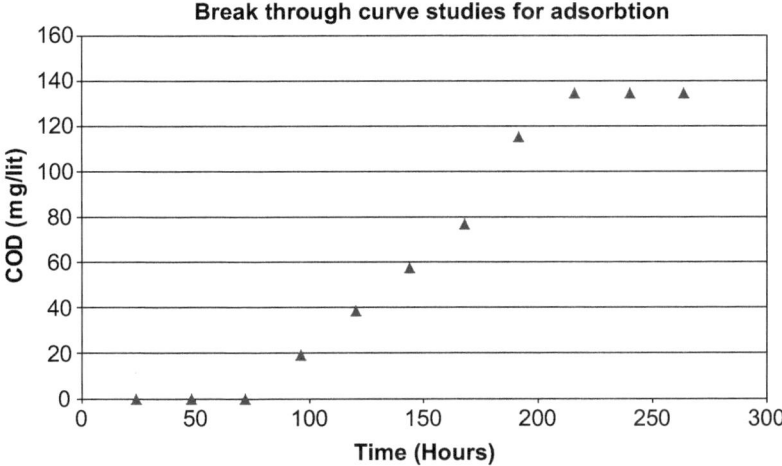

Fig. 1 Adsorption breakthrough data for bench-scale adsorber for COD removal flow rate = 10 mL/
min, column diameter = 30 mm, GAC (Commercial) 1.5 mm

Batch Equilibrium Experiments

Batch equilibrium experiments were undertaken using a series of fixed volume of
wastewater in jars, where they brought in contact with varying dosages of carbon.
The jars were placed in Phipps and bird multiple stirring device (Jar tester) apparatus and it was used at a constant temperature of 20 °C for 2 h until equilibrium was
reached. The experiments were conducted at a constant pH of 7.00. The optimal
dosage of carbon was found to be 5 g/L with a contact time of 60 min. The carbon
is then filtered off and the concentration remaining in solution was measured
through COD. The COD of the effluent was around 12–15 mg/L. The BOD of the
effluent was found to be below detection limit. The same sample was also analyzed
for other parameters such as TOC, TDS and heavy metals. The result of analysis
showed that, all the parameters are within the limits of discharge standards as prescribed by the Indian Standards (IS: 2490), and Ministry of Environment and Forests
(MOEF), Government of India.

6.5   Characterization of Sludges Generated from the Chemical
Treatment and Evaluation of Their Toxicity
Through Leaching Experiments

The sludge obtained due to chemical coagulation treatment was in grayish brown
colour. The pH of the sludge was around 1.40. As per the standard methods the
sludge volume index, weight of sludge generation, TCLP extraction, and heavy
metal analysis were done. The average analysis results of five sets of sludge samples
and TCLP samples are showed in Tables 9 and 10.

Table 9 Average analysis of sludge samples

Parameters	Units	Sludge samples average
Volume	ml/L	10.18 ± 0.308
Weight	mg/L	447 ± 11.72
Zinc	mg/kg	83.82 ± 54.08
Lead	mg/kg	0.35 ± 0.519
Nickel	mg/kg	0.75 ± 0.869
Cadmium	mg/kg	31.17 ± 30.87
Cobalt	mg/kg	BDL
Manganese	mg/kg	116.5 ± 75.63
Ferrous	mg/kg	938 ± 384
Chromium	mg/kg	80.52 ± 60.93
Copper	mg/kg	12.42 ± 6.60
Aluminium	mg/kg	174.75 ± 247.724

Table 10 Average analysis of TCLP samples

Parameters	Units	TCLP samples Average
Zinc	mg/L	BDL
Lead	mg/L	BDL
Nickel	mg/L	BDL
Cadmium	mg/L	BDL
Cobalt	mg/L	BDL
Manganese	mg/L	BDL
Ferrous	mg/L	BDL
Total Chromium	mg/L	BDL
Copper	mg/L	0.066 ± 0.023
Aluminium	mg/L	BDL

Physicochemical Treatment

The basis of the proposed physicochemical treatment plant design was made out from the wastewater outflow, characterization and treatability studies of the RL wastewater. The proposed scheme for the complete physicochemical treatment plant was shown in Fig. 2.

Flow Equalization

From the research laboratory wastewater outflow, the volume required for the flow rate equalization was determined by using inflow mass diagram. Based on that required volume, the flow equalization tank was designed to overcome the operational problems caused by the flow rate variations, to improve the performance of downstream processes, to improve the chemical feed control and process reliability, and to reduce the size and cost of the downstream treatment facilities.

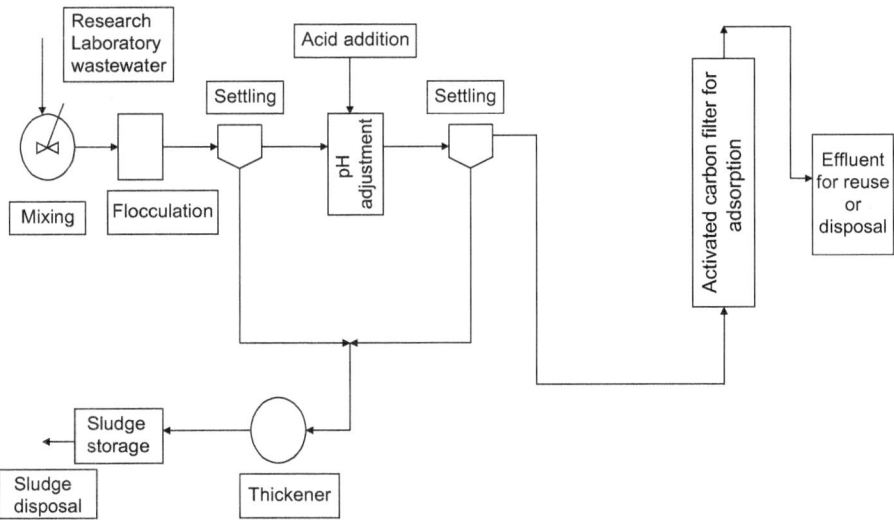

Fig. 2 Schematic presentation of the physicochemical treatment plant

7 Coagulation–Flocculation

Based on the chemical treatability studies experimental results, the design flow, reactor clarifier dimensions, surface loading rate, coagulant dosage requirements, and power requirements were calculated.

7.1 Adsorption

Based on the bench-scale and batch-scale experimental results, the design flow, dimensions for carbon column, carbon dosage requirements, power requirement calculations for backwash pump and piping were made.

7.2 Conclusion

The effluent generated after treatment with activated carbon had all the parameters within the limits of discharge standards as prescribed by the Indian Standards (IS: 2490). Thus, physicochemical treatment can be considered as a suitable treatment for the research laboratory effluent, and it is feasible to design an effective treatment plant and the treated effluent can be reused or safely discharged into public sewer/irrigation land/inland surface water bodies.

References

American Society for Testing and Materials (2000) Annual book of ASTM standards water and environmental technology. Practice for coagulation and flocculation jar test of water D2035-80. (ASTM, Easton, USA, Washington, DC, 1986)

USEPA (1986) Laboratory manual physical/chemical methods – test methods for evaluating solid waste SW –846 (office of the solid waste and energy response, USEPA, Washington, DC, 1986)

Efficiency of Various Granular Activated Carbon Adsorbents in the Treatment of Waste Water Generated from Research and Development Laboratories: A Case Study

Sundar Ramanathan and R. B. Lal

Abstract A host of chemicals/biochemicals is used in many research laboratories (RL) for R&D activities for various purposes. Such diverse R&D activities generate comparatively small quantities of waste streams with widely varying compositions. These waste streams may include aqueous solutions, organic solvents, heavy metals, and hazardous chemicals. As a result, the effluents generated in RL are highly toxic in nature, necessitating prior treatment before its discharge into the sewerage system.

In the present study, one of the RL effluents was characterized by the volume generated and chemical composition. Then a physicochemical treatment (PCT) sequence was evaluated with the objective of verifying the discharge standards in conformity with the Indian Standards.

In the PCT sequence, first the RL effluent was treated with lime and cationic polyelectrolyte sufloc-SN1 at a dosage of 400 and 2 mg L^{-1} respectively. After chemical treatment, the residual COD is 120.72 ± 13.97 mg L^{-1}. The other analyzed parameters are within the prescribed Norms. However, sometimes pollution load exceeds the prescribed Norms (IS: 2490-1982) due to the quantitative and qualitative variability of the analyses carried out in the RL, which results in different effluent compositions. So to reduce the residual COD and to overcome the future pollution load problems further treatment is necessary. The methods commonly employed for the destruction of residual COD are oxidation by means of ozone, hydrogen peroxide or adsorption by porous solids such as activated carbon, fly ash, and natural clays. Among these, Sorption on granular activated carbon (GAC) is one of the most widely used methods in water and wastewater treatments. GAC is considered one of the most effective adsorbents, especially for those substances

---

Note: The contents of this chapter indicates the author's view only and not that of the organization.

S. Ramanathan (✉)
Ministry of Environment, Forest and Climate Change, Regional Office, Chennai, India

R. B. Lal
Ministry of Environment, Forest and Climate Change, New Delhi, India

© The Author(s), under exclusive licence to Springer Nature Switzerland AG 2019   99
T. Jindal, *Emerging Issues in Ecology and Environmental Science*,
SpringerBriefs in Environmental Science, https://doi.org/10.1007/978-3-319-99398-0_8

containing refractory organic compounds that persist in the environment and resist biodegradation. In the adsorption process, the choice of GAC is justified by its good adsorbing capacity, due to the high surface area, resulting from the high porosity.

The objective of this study is to compare the adsorption efficiencies of two different GAC adsorbents with respect to the removal of residual TOC/COD from RL effluent. The adsorption efficiency for the two sets of GAC was determined by Freundlich adsorption model and breakthrough curve studies. Then the adsorbent with higher adsorption efficiency was chosen, and used in Bed depth service time model experiments to find the carbon bed efficiency. Further studies on chemical regeneration of chosen GAC was carried out to restore the maximum adsorption capacity and to retain as much as possible, the original pore structure of the adsorbent. Based on the experimental results, a full-scale adsorber design and treatment cost evaluation was done for both GAC adsorbents.

Keywords Lab · Waste water · Effluent · Activated carbon · Adsorption

1   Introduction

A host of chemicals/bio-chemicals are used in many research laboratory (RL) R&D activities for various purposes. Such diverse R&D activities generate comparatively small quantities of waste streams with widely varying compositions (Good win Stephanie 2003). These waste streams may include aqueous solutions, organic solvents, heavy metal elements and hazardous chemicals (National Research Council 1995). As a result, the effluents generated in RL are highly toxic in nature, necessitating prior treatment before its discharge in to the sewerage system.

In the present study, one of the RL effluents was characterized (APHA–AWWA–WEF 1998) by the volume produced and chemical composition. Then a physico-chemical treatment (PCT) sequence was evaluated with the objective of verifying the discharge standards in conformity with the Indian standards (Indian Standards: 2490-1982. Indian Standards for the Discharge of Industrial and Sewage Effluents, Government of India).

In the PCT sequence, first the RL effluent was treated with lime and cationic polyelectrolyte sufloc-SN1 at a dosage of 400 and 2 mg L^{-1} respectively. After chemical treatment, the residual COD is 120.72 ± 13.97 mg L^{-1}. The other analyzed parameters are within the prescribed norms. However, sometimes pollution load exceeds the prescribed Norms (IS: 2490-1982) due to the quantitative and qualitative variability of the analyses carried out in the RL, which results in different effluent compositions. So to reduce the residual COD and to overcome the future pollution load problems further treatment is necessary. The methods commonly employed for the destruction of residual COD are oxidation by means of ozone, hydrogen peroxide or adsorption by porous solids such as activated carbon, fly ash, and natural clays. Among these, sorption on granular activated carbon (GAC) is one of the most widely used methods in water and wastewater treatments (Hassler 1963). GAC is considered one of the most effective adsorbents, especially for those substances containing refractory organic compounds that persist in the environment and resist biodegradation (Abuzeid et al. 1995). In the adsorption process, the choice of GAC is justified by its good adsorbing capacity, due to the high surface area, resulting from the high porosity (Matson and Mark 1971).

The objective of this study is to compare the adsorption efficiencies of two different GAC adsorbents with respect to the removal of residual TOC/COD from RL effluent. The adsorption efficiency for the two sets of GAC was determined by Freundlich adsorption model and breakthrough curve studies. Then the adsorbent with higher adsorption efficiency was chosen, and used in Bed depth service time model experiments to find the carbon bed efficiency. Further studies on chemical regeneration (Toledo et al. 2003) of chosen GAC was carried out to restore the maximum adsorption capacity and to retain as much as possible, the original pore structure of the adsorbent. Based on the experimental results, a full-scale adsorber design and treatment cost evaluation were done for both GAC adsorbents.

2 Materials and Methods

2.1 Adsorbent Specifications

The adsorbents used in this study are denoted as AC-1 and AC-2, and they were manufactured by Merck KgaA, Darmstadt, Germany and Loba Chemie Pvt. Ltd, Mumbai, India, respectively. The chemical reagents used in this study are of analytical grade. Prior to adsorption experiments, both GACs were washed several times in de ionized water to remove all fine impurities, subsequently dried in an oven for one day at 110 °C, cooled at room temperature, for about 10 min and finally stored in air tight desiccators until they were used. The properties of AC-1 and AC-2 are given in Table 1.

2.2 Effluent Characterization

The RL effluent was first treated with lime and cationic polyelectrolyte (sufloc-SN1) at a dosage of 400 and 2 mg L^{-1} respectively. After their addition the pH and residual COD of the treated effluent was found to be 11.70 and 120.72 ± 13.97 mg L^{-1} respectively. The pH of the effluent was adjusted to neutral by adding specified

Table 1 Physical and Chemical properties of AC-1 and AC-2 GAC

Properties	Units	AC-1	AC-2
Total surface area (N_2 BET method)	m^2/g	190.56	916.89
Micro pore volume	cm^3/g	0.07620	0.3782
Adsorption average pore diameter	A°	16.8245	22.533
Bulk density	kg/m^3	400	490
Carbon	%	122.7988	129.8572
Nitrogen	%	9.8658	8.3605
Molecular formula	–	$C_{14}H_0N_1S_0$	$C_{18}H_0N_1S_0$

Table 2 Results of analysis of coagulant-treated wastewater

		Results of analysis of coagulant-treated wastewater		Indian standards for industrial and sewage effluents discharge IS: 2490-1982		
Parameters	Units	Influent (Raw waste water)	Effluent (Coagulant treated wastewater)	Inland surface	Public sewer	Land for irrigation
pH	–	7.58 ± 0.06	11.70 ± 0.1136	5.5–9.0	5.5–9.0	5.5–9.0
Colour	Pt-Co	38.2 ± 4.816	BDL	NS	NS	NS
Total dissolved solids	mg/L	1536.2 ± 133.288	1618.2 ± 52.36	2100	2100	NS
Total hardness	mg/L	331 ± 20.61	470 ± 7.16	NS	NS	NS
C.O.D.	mg/L	189.746 ± 20.61	120.72 ± 13.97	250	NS	NS
B.O.D.	mg/L	34 ± 8.21	17 ± 5.70	30	350	100
B.O.D./ C.O.D.	–	0.192 ± 0.072	0.143 ± 0.050	NS	NS	NS
Zinc	mg/L	0.229 ± 0.157	0.0022 ± 0.004	5.0	15.0	NS
Lead	mg/L	0.018 ± 0.017	BDL	0.1	1.0	NS
Nickel	mg/L	0.0022 ± 0.017	0.0011 ± 0.002	3.0	3.0	NS
Cadmium	mg/L	0.0132 ± 0.014	BDL	2.0	1.0	NS
Cobalt	mg/L	BDL	0.0052 ± 0.011	NS	NS	NS
Manganese	mg/L	0.024 ± 0.030	BDL	2.0	2.0	NS
Ferrous	mg/L	0.180 ± 0.114	BDL	3.0	3.0	NS
Total Chromium	mg/L	0.146 ± 0.024	0.0164 ± 0.028	2.0	2.0	NS
Copper	mg/L	0.020 ± 0.024	0.001 ± 0.002	3.0	3.0	NS
Aluminium	mg/L	0.340 ± 0.326	BDL	NS	NS	NS
Mercury	mg/L	0.003	BDL	0.01	0.01	NS

BDL below detection limit, *NS* not stipulate

amount of sulfuric acid solution. Prior to adsorption experiments, the treated effluent was thoroughly filtered by using what man No. 42 filter paper to remove the suspended particles. The characterization results of raw and treated effluent were given in Table 2. The average and standard deviation values of the characterization results were determined using Microsoft Excel software.

2.3 Equilibrium Isotherm Experiments

Batch equilibrium isotherm experiments were undertaken for AC-1 and AC-2 GAC by using a series of fixed volume of wastewater in jars, where known amounts of AC-1 and AC-2 GAC were added. The jars were placed in Phipps and bird multiple stirring device (Jar tester) apparatus and stirred at constant speed at 25 °C for 1 h. Then the supernatant was collected and analyzed for TOC parameter to determine the residual liquid concentration. This was then used to calculate the X/M ratio, by a material balance on the Freundlich adsorption system. The experiments were conducted at a constant pH of 7.

2.4 Lab-Scale Column Tests

The fixed bed column study was carried out to determine the breakthrough curve for the AC-1 GAC. The coagulant-treated effluent was fed to a feed tank, having a volume of 20 L and from which effluent was pumped using a peristatic pump at a constant flow rate of 25 ml/min. The effluent solution was fed through a bed of AC-1 GAC in down flow mode.

The carbon beds were contained in a glass column having a diameter of 30 mm and standard bed height of 300 mm. Approximately 100 g of AC-1 GAC was fed in to the column and samples were drawn at regular time interval from the column outlet. The collected samples were analyzed for COD parameter. The same procedure was followed for AC-2 GAC also.

2.5 Regeneration Experiments

The effect of regeneration on adsorption capacity of phenol was studied using AC-1GAC. Two successful regeneration cycles involving iterative adsorption and oxidation events were carried out. Adsorption was carried out by contacting the 50 mL (50 mg L^{-1}) of phenol solution with 0.5 g of AC-1 GAC in batch mode. (Sealed 250 ml borosilicate glass Erlenmeyer flask placed on an orbital shaker at 135 rpm, 2.5 days contact before oxidation.) In each adsorption event, pre- and post-aqueous solution samples were collected and analyzed to quantify the mass of phenol adsorbed. Following the adsorption event, the phenol loaded GAC was mixed with 10 ml of water, the pH was adjusted to 3 with sulfuric acid solution and the regeneration solution containing 11 mmol of hydrogen peroxide solution (30%), 0.27 mmol of ferrous sulfate hepta hydrate salt were added. The reaction flask was kept at room temperature for 2–3 h. After that, the reaction mixture was thoroughly filtered by using what man No. 42 filter paper and analyzed for phenol content in spectrophotometer as per the standard methods (APHA–AWWA–WEF 1998).

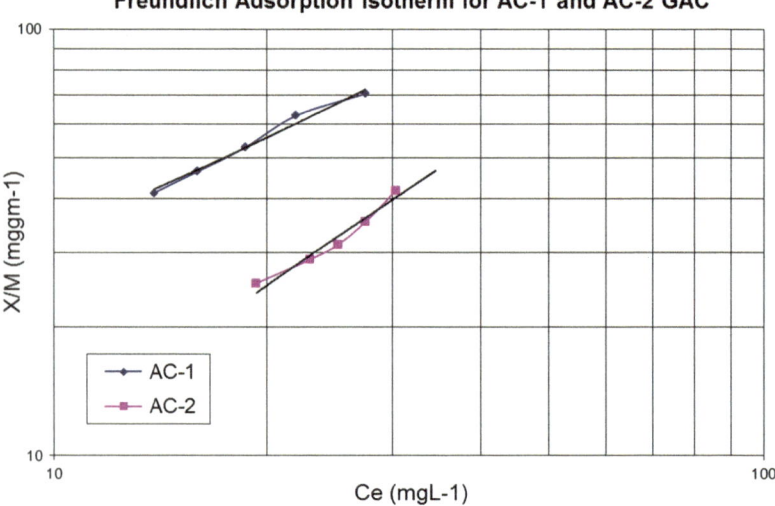

Fig. 1 Adsorption Isotherms for the adsorption of TOC by AC-1 and AC-2 GAC using the Freundlich adsorption isotherm model

2.6   COD Analysis

The chemical oxygen demand (COD) is an indication of over all oxygen load that a wastewater will impose on an effluent stream. COD is equal to the amount of dissolved oxygen that a sample will absorb from a hot acidic solution containing potassium dichromate and mercuric ions. The method adopted for the COD analysis was open reflux method (APHA–AWWA–WEF 1998). The COD of the coagulant-treated RL effluent is in the range of 120.72 ± 13.97 mg L^{-1}.

3   Results and Discussion

3.1   Equilibrium Isotherm Analysis

Data obtained from the equilibrium isotherm experiments are illustrated in Fig. 1 as a plot of residual TOC concentration (C_e) versus x/m. The experimental data were empirically fitted with the Freundlich adsorption isotherm model using Eq. (1).

$$x/m = kC_e^{1/n} \tag{1}$$

Where x is the amount of TOC removal, m, weight of carbon, x/m, concentration of TOC adsorbed per carbon dosage, C_e is the equilibrium concentration of TOC in

Table 3 Freundlich isotherm constants

GAC details	k	$1/n$	r^2
AC-1	11.363	0.451	0.9821
AC-2	4.551	0.675	0.9562

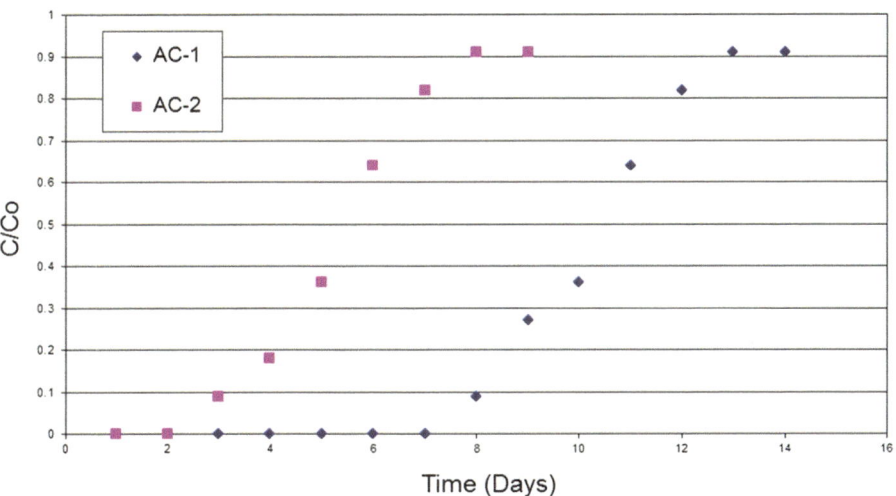

Fig. 2 Breakthrough curves for AC-1and AC-2 GAC in terms of service time for influent COD Concentration of 120.72 ± 13.97 mg L^{-1} ($D = 30$ mm, $H = 300$ mm)

solution, k and $1/n$ are constants. The graphical representations of this model are presented in Fig. 1. The respective constants k and $1/n$ are presented in Table 3.

Generally, a carbon that has a higher x/m value at a specified equilibrium concentration will be preferred for a given application (USEPA 1973). In Fig. 1, it appears that AC-1 had the higher x/m ratio at a given C_e value, while the AC-2 had the lower value.

The values of k and $1/n$ represent the intercept and slope of the Freundlich adsorption isotherms in Fig. 1. Larger values of k indicate the good adsorption capacity and the larger values of $1/n$ indicate the good adsorption intensity. Based on the k values only, adsorption capacity of AC-1 was nearly 2.5 times that of the AC-2 for the carbon dosages used in this study. The values of $1/n$ indicate that AC-2 had the higher adsorption intensity than AC-1. Therefore the AC-2 can adsorb organic components more rapidly than AC-1, it may have a more limited capacity for organic matter. The r-values for both GAC are greater than 0.95, it is assumed that the graphs on Fig. 1 are linear and Freundlich model is valid within the carbon dosage used.

3.2   Lab-Scale Column Tests

The experimental data obtained from the bench-scale adsorber are illustrated as breakthrough curves in Fig. 2. The results indicate that the performance of AC-1and AC-2 in removing residual COD from laboratory wastewater. The "S" shape curve is indicative of the effective use of adsorbents.

The carbon column was saturated after 7 and 2 days for AC-1 and AC-2 GAC, respectively. From the trend of the breakthrough curve, it was assumed that the process occurs in two phases. In the first phase a rapid diffusion of adsorbate takes place in the macro pore region; successively, during the final phase, a slower diffusion occurs in the region of micro pores, which could explain the saturation of the GAC.

3.3   Bed Depth Service Time Model

The BDST model is well established for fixed bed adsorption systems and therefore only a brief description is included. In the operation of fixed bed absorbers, the objective is to reduce the concentration in the effluent so that it does not exceed the predefined breakthrough value, C_b. Initially, when the activated carbon is unsaturated the actual effluent concentration is lower than C_b, but as the effluent is pumped through the bed the carbon becomes saturated and the effluent concentration approaches C_b, that is, the break point is reached. The original work on BDST model was carried out by Bohart and Adams (1920). He proposed a relationship between bed depth X, and time taken for breakthrough to occur. The service time t was related to the process conditions and operating parameters, see Eq. (2).

$$\ln\left(C_o / C_b - 1\right) = \ln\left(e^{K_a N_o X/v} - 1\right) - k_a C_o t \tag{2}$$

Hutchins (1973) proposed a linear relationship between the bed depth and service time, see Eq. (3).

$$t = N_o / C_o vX - 1/ k_a C_o \ln\left(C_o / C_b - 1\right) \tag{3}$$

$$t = a X + b \tag{4}$$

where a = Slope = $N_o/C_o \, v$; b = Intercept = $-1/k_a C_o \ln(C_o/C_b - 1)$.

If a value of a is determined for one flow rate, values for other flow rate can be computed by multiplying the original slope by the ratio of original and new flow rates. Accordingly, the BDST experiments were carried out for one flow rate and the slope was calculated. Then for another flow rate, the slope was computed through Hutchins procedure.

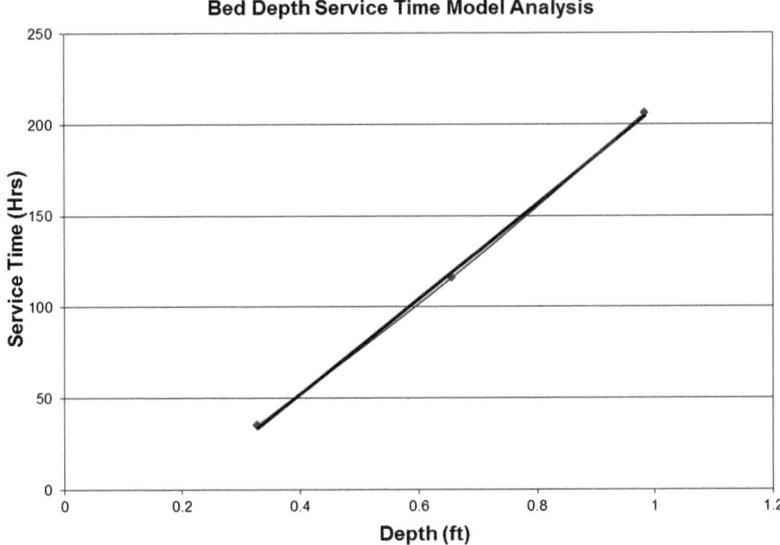

Fig. 3 Adsorption BDST plot for lab-scale adsorber for COD removal at a height of 0.10–0.30 m, column diameter = 0.03 m, column height = 0.30 m

Table 4 BDST experimental results

Hydraulic loading rate (m^3/m^2 min)	N_o (kg/m^3)	K_a (m^3/kg h)	X_o (m)	Bed efficiency
0.00846	91.68	0.271	0.060	79.6
0.0211	92.14	0.693	0.067	77.64

The analysis of the column experimental data using the BDST theory, yielded a plot of bed depth versus service time, illustrated in Fig. 3. The regression value was found to be 0.99.

The BDST parameters, namely adsorption capacity, N_o, rate constant k_a, critical bed depth X_o, and the bed efficiency were calculated from the experimental data and are presented in Table 4.

3.4   Effect of Regeneration

The effect of Fenton regeneration at room temperature for 2–3 h on adsorption capacity of phenol was studied using AC-1 GAC. After each regeneration cycle, the regeneration efficiency of AC-1 GAC was estimated based on the following formula.

$$\% \text{ Regeneration, Cycle } n = \frac{\text{Adsorption capacity after regeneration}}{\text{Adsorption capacity of fresh carbon}}$$

Table 5 Major design parameters for AC-1 and AC-2 adsorber

Parameters	Units	AC-1	AC-2
Diameter of the column	m	1.269	1.269
Height of the column	m	4.952	4.952
Volume of the column	m³	6.25	6.25
Mass of the carbon	kg	3057.6	2496
Life time of the carbon	days	63	12

Table 6 Treatment cost comparison for MGAC and LGAC

Details of the cost (Indian currency)	AC-1	AC-2
Carbon cost/year	1,200,000	1,562,500
Transportation cost/year	15,000	62,500
Labor cost/year	5000	25,000
Total cost/year	1,220,000	1,650,000
Net cavings/year	1,650,000 − 1,220,000 = 430,000	

The regeneration efficiency of first and second regeneration cycles was found to be 95% and 85.90% respectively. The reduction in regeneration efficiency may be due to the carbon deterioration (or) incomplete carbon regeneration.

4   Full-Scale Adsorber Design

Based on the lab-scale AC-1 column experimental results, scale-up study was carried out as per the standard design procedures (Wesley Ecken Felder Jr 2000). From that, the major design parameters for the plant scale adsorption column were determined. The same design procedure was followed for AC-2 also. The major design parameters for AC-1and AC-2 are given in Table 5.

5   Treatment Cost Comparison

The treatment cost comparison (Dawande 1992) studies were carried out for AC-1 and AC-2 by keeping the fixed capital cost, cost of the land and other non-depreciable facilities as a constant. In this comparison only the working capital, carbon transportation cost and the labor cost were considered.

From the plant scale adsorber design calculations, it was found that AC-1 requirement per adsorption column was 3 tonnes and its lifetime was 63 days. Similarly, the AC-2 requirement per adsorption column was 2.5 tonnes and its lifetime was 12 days.

If the adsorption column was operated 300 days in a year, then the no. of replacements for AC-1 and AC-2 were 5 and 25 times respectively. Depending upon the no. of replacements carbon cost, transportation and labor cost will vary. According to the current market rate, the costs of AC-1 and AC-2 are 80 and 25 rupees, respectively. Based on these details, treatment cost comparison studies was carried out and is given in Table 6.

From the treatment cost comparison studies, it is observed that AC-1 will be more effective and economical than AC-2. An annual savings of Rs. 430,000 can be obtained by using AC-1 as an adsorbent.

6 Conclusion

From the treatment cost comparison studies, it is observed that AC-1 will be more effective and economical than AC-2. An annual savings of Rs. 430,000 can be obtained by using AC-1 as an adsorbent.

References

Abuzeid N, Nakhla G, Farooq S, Oses-Twume E (1995) Activated carbon adsorption in oxidizing environments. Water Res 29:653–660

APHA–AWWA–WEF (1998) Standard methods for the examination of water and wastewater, 20th edn

Bohart GS, Adams EQ (1920) Adsorption in columns. J Chem Soc 42:523

Dawande SD (1992) Process equipment design. Central Techno Publications, Nagpur, India

Good win Stephanie (2003) Research highlights in waste treatment facility. A new decontamination plant at Lawrence Livermore National Laboratory, Science & Technology Review, pp 23–26, July/Aug 2003

Hassler JW (1963) Activated carbon. Chem. Publ., New York

Hutchins RA (1973) New methods simplifies the design of Activated Carbon system. Chem Eng 80(19):133–138

Matson P, Mark HB (1971) Activated carbon surface chemistry and adsorption from solution. Marcel Dekker, New York

National Research Council (1995) Prudent practices in the laboratory handling and disposal of chemicals. National Academic Press, Washington, DC

Toledo LC, Silva AC, Augusti R, Lago RM (2003) Application of Fenton's reagent to regenerate activated carbon saturated with Organo chloro compounds. Chemosphere 50:1049–1054

USEPA (1973) Process design manual for carbon adsorption. US Environmental Protection Agency, Technology Transfer Seminar Publication, Oct 1973

Wesley Ecken Felder W Jr (2000) Industrial water pollution control, 3rd edn. McGraw-Hill, New York

Index

© The Author(s), under exclusive licence to Springer Nature Switzerland AG 2019   111
T. Jindal, *Emerging Issues in Ecology and Environmental Science*,
SpringerBriefs in Environmental Science, https://doi.org/10.1007/978-3-319-99398-0